The Political Economy of Environmentally Related Taxes

OECD

ORGANISATION FOR ECONOMIC CO-OPERATION AND DEVELOPMENT

ORGANISATION FOR ECONOMIC CO-OPERATION AND DEVELOPMENT

The OECD is a unique forum where the governments of 30 democracies work together to address the economic, social and environmental challenges of globalisation. The OECD is also at the forefront of efforts to understand and to help governments respond to new developments and concerns, such as corporate governance, the information economy and the challenges of an ageing population. The Organisation provides a setting where governments can compare policy experiences, seek answers to common problems, identify good practice and work to co-ordinate domestic and international policies.

The OECD member countries are: Australia, Austria, Belgium, Canada, the Czech Republic, Denmark, Finland, France, Germany, Greece, Hungary, Iceland, Ireland, Italy, Japan, Korea, Luxembourg, Mexico, the Netherlands, New Zealand, Norway, Poland, Portugal, the Slovak Republic, Spain, Sweden, Switzerland, Turkey, the United Kingdom and the United States. The Commission of the European Communities takes part in the work of the OECD.

OECD Publishing disseminates widely the results of the Organisation's statistics gathering and research on economic, social and environmental issues, as well as the conventions, guidelines and standards agreed by its members.

This work is published on the responsibility of the Secretary-General of the OECD. The opinions expressed and arguments employed herein do not necessarily reflect the official views of the Organisation or of the governments of its member countries.

Also available in French under the title:
Économie politique et taxes liées à l'environnement

Foreword

Environmentally related taxes are increasingly used in OECD countries, and ample and increasing evidence of their environmental effectiveness is now available. However, there remains a high potential for a wider use of these environmental policy instruments, provided that they are well designed and that their potential impact on international competitiveness and income distribution are properly addressed. In particular, the environmental effectiveness and economic efficiency of environmentally related taxes could be improved further if existing exemptions and other special provisions were scaled back, and if the tax rates were better aligned with the magnitude of the negative environmental impacts to be addressed. Based on experience in OECD countries, this book gives a comprehensive discussion of these issues and of recent research on environmental and economic impacts of applying environmentally related taxes – in particular how to overcome obstacles to their implementation. It also discusses the environmental and economic effects of combining such taxes with other instruments for environmental policy.

The book is based on a number of in-depth case studies prepared for the Joint Meetings of Tax and Environment Experts under OECD's Committee on Fiscal Affairs and Environment Policy Committee. It has been prepared in close co-operation between Ana Cebreiro-Gómez, Chris Heady and Erik Vassnes of OECD's Centre for Tax Policy and Administration and Hope Ashiabor, Jean-Philippe Barde, Nils Axel Braathen and Pascale Scapecchi of OECD's Environment Directorate.

The book is published under the responsibility of the Secretary-General of the OECD.

Jeffrey Owens
Director,
Centre for Tax Policy and Administration,

Lorents G. Lorentsen
Director,
OECD Environment Directorate

Table of Contents

Figures

Organigram

ISBN 92-64-02552-9
The Political Economy of Environmentally Related Taxes
© OECD 2006

Executive Summary

Environmentally related taxes in the OECD countries

Experience over the last decades has proven that environmentally related taxes can be effective and efficient instruments for environmental policy. They introduce a price signal that helps ensure that polluters take into account the costs of pollution on the environment when they make production and consumption decisions. Many environmentally related taxes contribute to environmental improvements by causing price increases that reduce the demand for the environmentally harmful products in question.

All OECD member countries apply several environmentally related taxes. A database operated in co-operation between OECD and the European Environment Agency (EEA), currently details about 375 such taxes in OECD countries – plus i.a. some 250 environmentally related fees and charges. The taxes raise revenues in the order of 2-2.5% of GDP. About 90% of this revenue stem from taxes on motor vehicle fuels and motor vehicles, whereas revenue-raising is not a prime motivation for many other taxes applied.

The environmental effectiveness and economic efficiency of the environmentally related taxes could, however, be improved further if existing exemptions and other special provisions were scaled back. On the other hand, that could – to a certain extent – come in conflict with the two main political concerns that currently obstruct a more general use of tax rates that fully reflect the negative environmental impacts caused by various products and services, namely the fear of loss of sectoral competitiveness and the fear of negative distributional impacts. These two concerns are the focus of most of the present report.

Implementing environmentally related taxes

When implementing environmentally related taxes, the objectives of the policy measure should be clearly stated from the outset. When deciding a particular measure, one should carefully review the range of instruments that could potentially be used to achieve those objectives. A thorough analysis of the costs and benefits of each approach and an assessment of current practices should be carried out in order to evaluate the relative merits of the alternative measures. Environmentally related taxes can often usefully be implemented in the context of instrument mixes, in combination with other policy instruments, such as command and control regulations, tradable permits and – in certain cases – voluntary approaches.

International competitiveness

A major obstacle to the implementation of environmentally related taxes is often the fear of reduced international competitiveness in the most polluting, often energy intensive sectors of the economy. To date, environmentally related taxes have not been identified as causing significant reductions in the competitiveness of any sector. However, this is in part due to the fact that countries applying environmentally related taxes have provided for total or partial exemptions for energy intensive industries. Indeed, the OECD/EEA database (*www.oecd.org/env/policies/database*) shows that environmentally related taxes are levied almost exclusively on households and the transport sector. These exemptions create inefficiencies in pollution abatement and run contrary to the idea that polluters should pay.

With the Kyoto Protocol now in force, most OECD member countries have legally binding and quantified obligations to limit emissions of greenhouse gases. This has already contributed to new policy initiatives in several OECD member countries, with, for example, the *EU member States* implementing a CO_2 emissions trading scheme from 1 January 2005, with obligations for selected industries to hold emission allowances for the CO_2 emissions they cause.

Model simulations indicate that the use of economic instruments to reduce greenhouse gas emissions is likely to have negative impacts on the international competitiveness position of *some* industrial sectors, especially when such instruments are implemented in a non-global manner. This has, *e.g.* been demonstrated in case studies of the steel and the cement sectors. However, both studies show that in spite of some element of carbon leakage, significant global reductions in carbon emissions can be achieved.

Unilateral policies by single regions or countries may lead to significantly larger production decreases in the countries and sectors concerned. The larger the group of countries that put similar policies in place, the more limited are the impacts on sectoral competitiveness.

The case studies looked at some ways to limit the burden on affected firms, while maintaining the pollution abatement incentives. One option could be to recycle (a part of) the revenues raised back to the affected firms. The case studies indicate that revenue recycling would *reduce* global emission *reductions* in the sector. In other words, protecting competitiveness through recycling revenues back to the affected sectors is likely to lower the environmental effectiveness of the policy as a whole. The report also discusses the use of border tax adjustments to reduce competitiveness impacts of taxes.

This report also describes some *ex post* case studies of the implementation of environmentally related taxes, highlighting lessons of political economy that can be drawn. One lesson is that policy makers should ensure that competitiveness pressures are adequately assessed and addressed. Countries should also strive for broadest possible tax-bases to ensure cost-efficient emission reductions. Broad tax-bases and introduction in connexion with a broader reform strategy might make it easier to get acceptance for the tax from affected parties. This strategy has been used in many countries that have introduced green tax reforms.

Income distribution

The distributional incidence of environmental policy has become a key issue in the policy debate. Most studies show that environmentally related taxes, and especially energy taxes, can have a regressive impact on the income distribution of households. However, a full assessment of the income distributional effects of levying environmentally related taxes should also include the indirect effects from price increases on taxed products, effects arising from the use of environmental tax revenues and/or compensational measures and also the distribution of the environmental benefits resulting from the tax.

Mitigation practices, including exemptions or rate reductions in the environmentally related taxes, can reduce the environmental effectiveness of the taxes. Governments should seek other, and more direct, measures if low-income households are to be compensated. Compensation measures, through reductions in other taxes or through the social security system, can maintain the price signal of the tax whilst reducing the negative impact of the tax on low-income households.

Regressive impacts from implementing environmental taxes are often softened by using the revenue to reduce other taxes i.a. on income. Such tax reductions can be targeted at low-income groups. In some cases the distributional concerns have not been addressed at all, or have come up late in the process and tackled in a more *ad hoc* fashion. This might lead to strong opposition and failure to implement effective environmental measures, and implies higher costs to society than necessary. In order to assure that distributional concerns are properly addressed, member countries should consider introducing mechanisms into the decision-making process whereby distributional impacts are explicitly analysed.

The use of tax revenue

Revenue from introducing environmental taxes can be used to strengthen the budget balance, finance increased spending or reduce other taxes. Several OECD member countries have reduced the tax burden on labour, by cutting non-wage labour cost in the form of employers' social security contributions. This can reduce the efficiency loss generally induced by tax collection if the taxes reduced are more distorting than the environmentally related taxes introduced. One much-debated aspect of use of revenue from environmental taxes is shifting the tax burden from labour to pollution with an expectation that this would encourage work effort and thereby contribute to increased employment, while improving the environment (the so-called "double dividend" hypothesis). A distinction is often made between a strong and a weak double dividend hypothesis. The weak double dividend hypothesis states that revenue recycling through cuts in distortionary taxes improves welfare relative to recycling through lump-sum payments. The strong double dividend suggests that substitution of an environmental tax for a representative distortionary tax will improve welfare. The weak double dividend hypothesis is not disputed, but the strong one is.

Tax revenues are sometimes earmarked to specific spending purposes, in some cases for environmental purposes. Earmarking, however, might violate the polluter pays principle. Further, earmarking fixes the use of revenue which may create an obstacle for the re-evaluation and modifications of the tax and spending programs. Therefore, the economic and environmental rationale for any earmarking should be evaluated regularly to avoid inefficient spending that would otherwise not be financed from general tax revenues.

Administrative costs

Environmentally related taxes can be designed in ways that imply relatively low administrative costs. For example, taxes on petroleum products are levied on a limited number of petroleum refineries and depots, and are relatively simple to administer and enforce. Several examples also indicate that the administrative costs of a scheme involving a large number of tax payers *can* be kept at relatively modest levels.

However, many economic instruments used for environmental policy involve several special provisions that increase the administrative costs. Such mechanisms, like for instance exemptions, are often introduced for non-environmental reasons, to address competitiveness or income distribution concerns. Often there seems to be a trade-off between the size of the administrative costs and measures to create a "fair" or "politically acceptable" scheme. Thus, the design features of environmentally related taxes associated with low administrative costs are often in line with features associated with high economic efficiency.

Acceptance building

Building acceptance is a key condition for successful introduction of environmentally related taxes. The acceptance of an economic instrument among the public at large seems to be related to the degree of awareness of the environmental problem the instrument seeks to address. In general, political acceptance could be strengthened by creating a common understanding of the problem at hand, its causes, its impacts, and the impacts of possible instruments that could be used to address the problem. One way to build such a common understanding is to provide correct and targeted information on the environmental issues at stake. Another way is to involve relevant stakeholders in policy formulation, for example through broad formal consultations and/or in committees or working parties preparing new policy instruments.

Environmentally related taxes used in combination with other instruments

In practice, environmentally related taxes are seldom used in complete isolation. Taxes are for instance often applied in combination with regulatory instruments. In a number of cases there can be environmental and/or economic benefits from combining a tax with other types of policy instruments. A tax can relatively well affect the total amount used of a given type of product and the choice between different product varieties, but could – *inter alia* for monitoring and enforcement reasons – be less suited to address *how* a given product is used, *when* it is used, *where* it is used, etc. Hence, other instruments could in any case be needed.

On the other hand, it often seems that more environmental targets than necessary have been defined. This could be the case in the waste area, where *e.g.* specific recycling targets for a large number of products or waste streams have been established in many OECD countries, frequently without a clear documentation that the selected waste streams represent a larger threat to the environment than other waste streams. Also the targets set for landfill diversion of biodegradable waste could benefit from further cost-benefit analyses.

ISBN 92-64-02552-9
The Political Economy of Environmentally Related Taxes
© OECD 2006

Chapter 1

Introduction, Background and Main Findings

1.1. Introduction and background

Experience over the last decades has proven that environmentally related taxes can be effective and efficient instruments for environmental policy. The environmental effectiveness and economic efficiency of the environmentally related taxes applied in OECD member countries could, however, be improved further if existing exemptions and other special provisions included in the taxes were scaled back, and if the tax rates were better aligned with the magnitude of the negative environmental impacts to be addressed.

OECD last issued a publication on environmentally related taxes in 2001, see OECD (2001a). Since then a lot of additional work have been undertaken under the auspices of the Joint Meetings of Tax and Environment Experts under the Committee of Fiscal Affairs and the Environment Policy Committee, in particular on how to address concerns about negative impacts of such taxes on sectoral competitiveness, on income distribution and on the political economy of their introduction. The present publication builds mostly on this work, but also incorporates recent work by other authors.

1.2. Current use of environmentally related taxes

As described in detail in Chapter 2, a large and growing number of environmentally related taxes are applied in OECD member countries – as well as a vast number of environmentally related fees and charges.[1] They raise revenues in the order of some 2-2.5% of GDP. About 90% of the revenue is raised through taxes on motor vehicle fuels and motor vehicles, whereas revenue-raising is not the prime motivation for many of the other taxes applied.

The environmentally related taxes in OECD countries include more than 1 150 exemptions and several hundred refund mechanisms and other tax provisions, favouring various economic sectors, various products and/or various uses of certain products. These provisions are introduced for a number of different reasons, such as to limit any negative impacts on the international competitiveness of some economic sectors, to alleviate economic hardship for certain households or to promote the use of more environmentally benign product varieties. Except for the latter case, such tax provisions will generally tend to reduce the environmental effectiveness of the taxes – and reduce the economic efficiency with which environmental policy targets are met.

1.3. Environmental effectiveness

A first point to underline is that *any* policy instrument used to achieve environmental targets *should* cause changes in consumption and/or production patterns. If an instrument fails to create such changes, it simply cannot deliver any environmental improvements. The more relevant issues are, hence, *who* should change their behaviour, by *how much* and within which *timeframe*.[2]

The information provided in Chapter 3 clearly demonstrates that many existing environmentally related taxes – and environmentally motivated tax rate variations in some of these taxes – *are* contributing to environmental improvements. Whereas the demand for many of the relevant tax-bases in economic terms would be described as "inelastic" – since their own-price elasticities are smaller than 1 in absolute value – it is clearly documented that most of the price elasticities are significantly different from 0. A price increase (triggered by a tax increase or by some other factor) *will* therefore reduce the demand for the product in question.

The environmental effectiveness of many of the environmentally related taxes used today could, however, be increased even further if existing exemptions and other specific tax provisions were scaled back. On the other hand, that could – to a certain extent – come in conflict with the two main political concerns that currently tend to obstruct a more general use of tax rates that fully reflect the negative environmental impacts caused in the prices of various products and services (also called "internalising the environmental externalities"), namely the fear of loss of sectoral competitiveness and the fear of negative distributional impacts. These two concerns are the focus of most of the present report.

1.4. Sectoral competitiveness

It is important to distinguish between competitiveness impacts at a *national* and at a *sector* or *firm* level. Policy changes that make some firms worse off will *always* make other firms better off. At a national level any negative impact imposed upon one firm or sector will thus tend to be attenuated by positive impacts on others.

While there are many good reasons why policy makers *ought to* focus more upon impacts of (environmental) policies at a national than at a sectoral level, *in practise* they tend to be more concerned with any potential "losers" from a policy change than with the impacts on the economy as a whole.[3] Hence, the focus discussion in this publication is on competitiveness impacts at a sectoral level – not at a national level. Considerable attention is, by the way, given to energy policies and policies addressing climate change, in part because such policies are likely to trigger more significant behavioural changes throughout the economy than most other environmental policies.

With the Kyoto Protocol now in force, most OECD member countries have legally binding and quantified obligations to limit emissions of greenhouse gases. This has already contributed to new policy initiatives in several OECD member countries, with, for example, the EU member States implementing a CO_2 emissions trading scheme from 1 January 2005, with obligations for the power sector and selected industries to hold emission allowances for the CO_2 emissions they cause.

Model simulations indicate that the use of economic instruments to significantly reduce greenhouse gas emissions is likely to have negative impacts on the international competitiveness position of *some* industrial sectors, especially when such instruments are implemented in a non-global manner. This has, *e.g.* been demonstrated in recent OECD case studies of the steel and the cement sectors.[4]

It can in any case be useful to keep in mind that the closing down of (otherwise) unprofitable and highly-polluting firms could be the lowest-cost way for society to reach a given environmental target, not withstanding the social implications of doing so. For example, many steel firms have been depending on significant subsidies for the last decades. OECD is working to reduce subsidies to the steel sector in particular[5] and environmentally harmful subsidies in general, see OECD (2003e and 2005c).

The steel case study indicated *inter alia* that an *OECD-wide* carbon tax of 25 USD per tonne CO_2 would reduce OECD steel production in the order of –9%. The estimated reduction was much greater for the heavily polluting integrated steel mills (–12%) than for the scrap-based mini-mills (–2%). Non-OECD production would increase by almost 5%, implying a fall in world steel production of –2%. The carbon tax would induce some substitution from the use of pig iron towards more intensive use of scrap in Basic Oxygen Furnace (BOF) steel making. Scrap prices would then rise, thus weakening the competitiveness of scrap-based Electric Arc Furnace (EAF) steel producers.

A first lesson that can be drawn is that different firms within a given sector will *not* be affected in the same way by any use of economic instruments, due to the different input combinations and the resulting differences in emission profiles.

A second lesson is the importance of taking into account possible adjustments in *related markets* when considering the impacts of a given policy on a particular sector. A part of any initial burden placed on a sector is likely to be shifted backwards to input suppliers and forward to the customers. In the steel case, this is *e.g.* illustrated by the estimated impacts on scrap metal prices, and the increase in steel prices.

The estimated OECD-wide tax would reduce OECD emissions of CO_2 from the steel industry by 19%. Despite relatively high emission intensities in non-OECD countries, global emissions from the sector would decline by 4.6%, *i.e.* more than twice the reduction in global steel production. This is due to substitution towards a cleaner input mix and cleaner processes in the OECD area.

A similar finding is made in the cement case study. In a scenario with emission trading in Annex B countries except USA and Australia, even if cement production outside this area increases in response to the policies put in place in most of the OECD-related regions, global carbon emissions in clinker manufacturing is estimated to decrease by some 32 million tonnes carbon in 2010 (around 2% in 2010, 2020 and 2030).

A third lesson is thus that, in spite of *some* element of "carbon leakage" also when policies to combat climate change are put in place on a relatively broad front, significant global reductions in carbon emissions – compared to a reference scenario – can be achieved.[6]

Unilateral policies by single regions or countries may lead to significantly larger production decreases in the countries and sectors concerned. A fourth lesson is that – while the importance varies between different firms – the larger the group of countries that put similar policies in place, the more limited the impacts on sectoral competitiveness.

The case studies looked at some potential ways the burden on affected firms could be limited, while maintaining their abatement incentives. One option could be to recycle (a part of) the revenues raised back to the firms in question. The studies reinforce the first lesson drawn above, that different firms within a given sector will be affected in different ways by any use of economic instruments.

Another important point is that revenue recycling would *reduce* global emission *reductions* in the sector. In other words, protecting the competitiveness of energy-intensive sectors in the OECD area through the recycling of tax revenues to the given sectors is likely to lower the environmental effectiveness of the policy as a whole. Simulations show that recycling tax revenues while contracting OECD production only by around 1% (instead of 9%), will reduce global emissions by around 3% (in contrast to 4.6%).

So-called "Border Tax Adjustments" represent another option to limit sectoral competitiveness impacts of economic instruments. From an environmental point of view there could, in some cases, be advantages related to the use of border tax adjustments. However, both practical and legal issues related to their implementation need to be solved. For a further discussion, see Chapter 5 below.

Lessons on political economy can also be drawn by analysing *ex post* case studies on the implementation of instruments with potential negative impacts on competitiveness. We analysed some empirical country case studies: the proposed industrial energy consumption tax in France, the United Kingdom Climate Change Levy, the MINAS in the Netherlands, the Swiss heavy goods vehicle road use fee, the Irish plastic bag tax and the Norwegian aviation fuel tax.

A first lesson from these *ex post* case studies is that policy makers should take steps to ensure that competitiveness pressures are adequately assessed and addressed. In doing so, it is important to consider the mitigation measures against any legal obligations and to ensure that the measure will not be found to provide a prohibited subsidy (*e.g.* the proposed industrial energy consumption tax in France).

A second lesson is that, as also seen in the theoretical case studies, when loss of competitiveness is an issue different mitigating measures can be considered and they will have different effects on both environment and competitiveness. When considering different measures it is important that they do not reduce abatement incentives. When levying taxes that raise revenue, many countries have used compensational measures by reducing other taxes (for instance as in case of the Norwegian aviation fuel tax) or other kinds of budgetary compensation. Some countries have introduced sectoral exemptions or reduced rates (as for instance is the case in the UK Climate Change Levy). Finally, sometimes international co-ordination at different levels can be useful and even necessary for implementing market-based instruments addressing environmental problems (as in the case of the Swiss heavy vehicle fee, where the bilateral agreement with the EU was important for the implementation).

However, one should note that there often seems to be a trade-off between the size of the administrative costs and measures to create a "fair" or "politically acceptable" scheme. Often mechanisms introduced for non-environmental reasons, to address competitiveness or income distribution concerns, are responsible for an increase of the administrative costs, *e.g.* CCL in the UK and the MINAS nutrients accounting system in the Netherlands.

Additionally, relatively modest compensation mechanisms can often suffice when introducing a tax or a trading scheme (even based on auctioning), in order to make the owners of the firms equally well-off as before – but the size of the "necessary" compensation depends on how insulated the domestic market is from international competition. However, there is a risk that the affected firms could be seriously over-compensated. If so, the economic efficiency costs will increase because, for example, less money would be available to reduce distortionary taxes.

The Swiss case is also a good example of the importance of seizing the right moment for pushing through a delicate project on the political agenda. Therefore, a fourth lesson is that a project that at some point in time seems impossible to implement might be feasible when the circumstances are more favourable.

A fifth lesson is that countries should strive for broadest possible tax-bases to ensure cost-efficient emission reductions. Broad tax-bases and introduction in connection with a broader fiscal reform strategy might make it somewhat easier to get acceptance for the tax from affected parties and thus might contribute to a smooth implementation. This strategy also seems to have been followed in many countries that have introduced green tax reforms.

The case study of the Irish plastic bag tax shows the importance of doing thorough initial research and carefully considering other relevant policy options. Introducing a tax is not always the right answer. This study assessed several policy options/instruments to address the environmental problems created by plastic bags in Ireland. Therefore, a sixth lesson is that one should consider other measures, in addition to environmentally related taxes, to tackle an environmental problem.

Finally, based on the case of the Swiss heavy vehicle fee, one can also draw the lesson that a gradual phasing in of taxes can soften the immediate cost impact and give companies time to adjust to reduce the tax burden.

1.5. Income distribution

Most studies show that the direct effects of environmentally related taxes, and especially energy taxes, can have a regressive impact on the income distribution of households. However, empirical analysis indicate that the degree of regressivity decreases once the indirect distributional effects from price increases on taxed products and the environmental effects of the tax are taken into account. Further, when taking account of measures of mitigation or compensation, the regressive impact of environmentally related taxation can in most cases be softened and even removed. Then the net effect of the environmental policy can even end up being progressive.

A full assessment of the income distributional effects of levying environmental taxes should also include indirect distributional effects from price increases on taxed products, effects arising from the use of environmental tax revenues and/or compensational measures and also the distribution of the environmental benefits resulting from the tax.

Mitigation practices reduce the environmental effectiveness of taxes. In the case of regressivity, governments should seek other, and more direct, measures if lower-income households are to be compensated. Such compensation measures can maintain the price signal of the tax whilst reducing the negative impact of the tax on household income. A key policy consideration is to maintain abatement incentives for the households in question. The incentive to consume or use less of a given product are created by the (assumed) imposition of environmental taxes, which drive up purchase prices to reflect the social damage created by their consumption or use.

Undesirable distribution effects can in general be addressed through the *social security systems* and *tax systems*. Relief from an environmentally related tax through a personal income tax system can *i.a.* include: increases in a basic personal allowance, introduction of non-wastable or wastable tax credits. Wastable tax credits are attractive, relative to tax allowances, because they avoid inter-actions with the tax rate structure. However, wastable tax credits do not deliver in full the intended amount of tax relief where an individual has insufficient income to fully absorb the tax credit. Disregarding any budgetary concerns, non-wastable tax credits might be preferred, because they provide *cash transfers* for credit amounts that cannot be used to offset personal income tax liabilities.

Where it is necessary to alleviate the tax burden on certain groups, such concerns should be addressed, to the extent possible, by approaches unrelated to current consumption decisions. In other words, it is important that relief is not provided through exemptions to the environmentally related taxes themselves, or through reduced tax rates on consumption for targeted groups. Aside from the possible need to provide compensation to win political support for a tax, equity and efficiency goals may be better addressed by explicitly targeting the relief to the more vulnerable and needy households.

Experiences from some member countries show that regressive impacts from implementing environmental taxes are often softened by using the revenue to reduce other taxes i.a. on income. Then the tax reductions can be targeted at lower income groups. In other cases the distributional concerns have not been addressed at all, or have come up late in the process and been tackled in a more *ad hoc* fashion. This might lead to large opposition and failure to implement effective environmental measures and implies higher costs to society than necessary.

In order to assure that distributional concerns are properly addressed, member countries should consider introducing measures that implement considerations of distributional concerns into the decision-making process. Some countries have therefore introduced specific *institutional arrangements,* as for instance specialised working groups or committees. Other countries have developed specific *guidance documents* for policy makers.

1.6. Administrative costs

It is *possible* to design a number of economic instruments for environmental policy with relatively low administrative costs. For example, taxes on petroleum products are usually levied on a limited number of petroleum refineries and depots, and are hence relatively simple to administer and enforce. For instance, the administrative costs of the ecological tax reform in Germany are estimated to comprise just 0.13% of the revenues raised. Several examples also indicate that the administrative costs of a scheme involving a large number of tax payers *can* be kept at relatively modest levels.

However, many economic instruments used for environmental policy do involve a large number of special provisions that increase the administrative costs. Such mechanisms are often introduced for non-environmental reasons, to address competitiveness or income distribution concerns. A lesson that can be drawn is that there often seems to be a trade-off between the size of the administrative costs and measures to create a "fair" or "politically acceptable" scheme.

1.7. Political acceptance

A first point to make is that the "acceptance" of an economic instrument among the public at large seems to be related to the degree of awareness of the environmental problem the instrument is to address, and whether this instrument is considered to contribute significantly to reducing the environmental problem. A policy-implication of this is that it is advisable to "prepare the ground" for later instrument implementation by providing correct and targeted information to the public on the causes and impacts of relevant environmental problems.

It also seems clear that the degree of political acceptance depend on the perceived "fairness" of the instrument in question. A lot of the attention concerning "fairness" is related to perceived sectoral competitiveness impacts and/or impacts on low-income households. Policy makers could do well in trying to shift the discussion of "fairness" more towards "who are the most important contributors to the problem at hand" – and to try to make people more aware that *any* instrument that could be applied to address a given problem will have (positive or negative) income distribution impacts.

In general, political acceptance could be strengthened by creating a common understanding of the problem at hand, its causes, its impacts, and the impacts of possible instruments that could be used to address the problem. One way to build such a common understanding is to involve relevant "stakeholders" in policy formulation, for example through broad formal consultations and/or in committees or working parties preparing new policy instruments.

1.8. Environmentally related taxes used in combination with other instruments

In a number of cases there can be environmental and/or economic benefits from combining a tax with other types of policy instruments. In practice, environmentally related taxes are seldom used in complete isolation – in a large number of cases one or more regulatory instruments will, *for example*, be applied. The mere existence of instrument mixes is, however, obviously not a "proof" of their environmental effectiveness and economic efficiency.

A rather obvious first requirement for applying an environmentally efficient and economically effective instrument mix is to have a good understanding of the environmental issue to be addressed. In practice, many environmental issues can be more complex than perhaps first thought, as they often have a number of *relevant*, and often *correlated*, "*aspects*" or "*characteristics*" – and many of the instruments that are applied contain a large number of separate "*rules*" or "*mechanisms*". A tax (or a tradable permit system) can relatively well affect the total amount used of a given type of product and the choice between different product varieties, but could – *inter alia* for monitoring and enforcement reasons – be less suited to address, for example, *how* a given product is used, *when* it is used, *where* it is used, etc. Hence, other instruments could in any case be needed.

On the other hand, in some cases it can seem that more environmental targets than necessary have been defined. This could be the case in the waste area, where – for example – specific recycling targets for a large number of products or waste streams (*e.g.* packaging) have been established in many OECD countries, frequently without a clear documentation that the selected waste streams represent a larger threat to the environment than other, related, waste streams. Also the targets set for landfill diversion of biodegradable waste could benefit from further cost-benefit analyses.

A second requirement for designing efficient and effective policies is to have a good understanding of the links with other policy areas. In addition to co-ordinating different environmental policies, co-ordination with other related policies – such as energy policies, housing policies, agricultural policies, transport policies, etc. – is needed.

A third requirement is to have a good understanding of the interactions between the different instruments in the mix. Various instruments can interact with environmentally related taxes in a number of ways. For example:

- A labelling system can help increase the effectiveness of a tax by *providing better information* to the users on relevant characteristics of different product the tax applies to. The price elasticities of concern can hence increase.

- Combining a tax on energy use with targeted subsidies for better isolation of buildings can be a good way to *address split incentives* between landlords and tenants.

- The combination of a tax and a voluntary approach, like *e.g.* a negotiated environmental agreement, can *increase the "political acceptability"* of the former – by limiting any negative impacts on sectoral competitiveness – at the cost of reduced environmental effectiveness or increased economic burdens placed on other economic actors.

- Combining a tax and a tradable permits system can help *limit compliance cost uncertainty* – compared to the application of a trading system in isolation.

- On the other hand, such a combination would *increase the uncertainty related to the environmental effectiveness*.

- There is also a danger that a regulatory instrument applied next to an environmentally related tax could *unnecessarily restrain the flexibility* for polluters to find cost-effective abatement options offered by a tax.

Notes

1. Definitions of environmentally related taxes, fees and charges are given in Box 2.1 in Chapter 2.

2. In addition, it is a relevant issue whether or not given environmental targets – or the lack of such targets – represent a reasonable balance between the benefits and costs of environmental improvements.
   The choice of policy instruments can, however, affect the efficiency at which a given target it reached – and affect the rate of new technology developments, which can be of importance for the cost to society of reaching given policy targets in the longer term. Economic instruments, like taxes or tradable permits, *can* help in achieving a given target at the lowest possible cost to society as a whole, both in the short term, as they can equalise marginal abatement costs between polluters (obtain *static efficiency*), and in the long term, as they provide a continuous incentive for further technology development.

3. Even if there *are* reasons to provide (temporary) relief to those that lose out from a policy change, it is by no means given that this best can be done through some modification to the tax in question.

4. See OECD (2003d and 2005f).

5. See *www.oecd.org/document/5/0,2340,en_2649_34221_32362885_1_1_1_1,00.html*.

6. The word "significant" is of course relative. The emission reductions obtained in the simulations discussed here are small compared to what would be needed to fulfil the long-term objectives of the UN Framework Convention on Climate Change.

ISBN 92-64-02552-9
The Political Economy of Environmentally Related Taxes
© OECD 2006

Chapter 2

Current Use of Environmentally Related Taxes

All OECD member countries apply several environmentally related taxes – according to the definition agreed between OECD, IEA and the European Commission (see Box 2.1). A database operated in co-operation between OECD and the European Environment Agency (EEA) currently details about 375 such *taxes* in OECD member countries – plus some 250 environmentally related fees and charges in the same countries.[1] This chapter draws extensively on that database to provide an overview over the taxes that are applied, the amount of revenues raised, the tax-bases on which the taxes are levied, the tax rates implemented, and of the exemptions, refund mechanisms, etc. included in the taxes.[2]

Box 2.1. **Defining environmentally related taxes**

OECD, IEA and the European Commission have agreed to define environmentally related *taxes* as any compulsory, *unrequited* payment to general government levied on tax-bases deemed to be of particular environmental relevance. The relevant tax-bases include energy products, motor vehicles, waste, measured or estimated emissions, natural resources, etc. Taxes are unrequited in the sense that benefits provided by government to taxpayers are not normally in proportion to their payments.

Requited compulsory payments to the government that are levied more or less in proportion to services provided (*e.g.* the amount of wastes collected and treated) can be labelled as *fees* and *charges*. The term *levy* covers both taxes and fees/charges.

2.1. The taxes applied

The largest number of environmentally related *taxes* in OECD countries is levied on energy products (150 taxes) and on motor vehicles (125 taxes). There is also a significant number of waste-related *taxes* in OECD (about 50 taxes in all), levied either on certain products that can cause particular problems for waste management (about 35 taxes), or on various forms of final waste disposal, *i.e.* on incineration and/or landfilling (15 taxes in all).[3] The remaining 40 or so environmentally related taxes applied in OECD countries are levied on a broad spectre of tax-bases, as can be seen from Table 2.1, that details these *taxes* and the waste-related *taxes* included in the database. Among the taxes relatively recently introduced in some countries are taxes on various hazardous chemicals and on the extraction of certain natural resources, like sand and gravel.

The definition of environmentally related taxes does *not* focus on the (alleged) purpose of a tax, or on how the revenues raised by the tax are used. However, the revenues of about 1/3 of the taxes in OECD countries *are* earmarked for a particular purpose – often not for an environmental purpose per se. About 75 of the earmarked taxes are levied on energy products (including 50 motor fuels taxes levied at state level in the United States), 15 are levied on motor vehicles while 20 are waste-related taxes. Whereas the earmarked transport-related taxes (including the taxes on motor fuels) tend to be allocated to the construction or maintenance of roads, etc., earmarked waste-related taxes are normally

Table 2.1. **Environmentally related taxes in OECD countries not levied on energy or transport 1.1.05**

AUSTRALIA	Duty on nitrogen	Tax on groundwater extraction
New South Wales – Waste levy	Duty on pesticides	Tax on tap water
Oil recycling levy	Duty on waste	Waste tax
Aircraft noise levy	Duty on sealed NiCd-batteries	Tax on the pollution of surface waters
Ozone protection and synthetic greenhouse gas levy	Duty on carrier bags made of paper, plastics, etc.	
	Duty on tyres	**NORWAY**
AUSTRIA	Duty on polyvinyl chloride and phthalates	Tax on final treatment of waste
Waste deposit levy	Duty on waste water	Product tax on beverage containers
	Excise duty on antibiotics and growth promoters	Basic tax on non-refillable beverage containers
BELGIUM	Tax on water quantity	Tax on lubricating oil
Environmental taxes		Tax on pesticides
Packaging charge	**FINLAND**	Tax on trichloroethane and tetrachloroethane
Flanders – Groundwater tax	Oil damage levy	
Flanders – Tax on waste dumping and burning	Oil waste levy	**SPAIN**
Flanders – Manure tax	Excise on disposable beverage containers	Andalusia – Tax on emissions to air
Flanders – Water pollution tax	Tax on waste	Andalusia – Tax on radioactive waste
Wallonia – Tax on waste collection		Andalusia – Tax on coastal wastewater discharge
Wallonia – Tax on water withdrawals	**FRANCE**	Galicia – Tax on emissions to air
	General tax on polluting activities	
CANADA (*Examples only*)		**SWEDEN**
British Columbia – Batteries tax	**HUNGARY**	Tax on waste
British Columbia – Logging tax	Air pollution levy	Natural gravel tax
British Columbia – Mining tax	Noise abatement levy	Tax on pesticides and artificial fertilisers
British Columbia – Tyres tax	Product charge on packaging materials	
British Columbia – Tax on lead acid batteries	Toxic waste levy	**SWITZERLAND**
Federal air conditioner tax	Product charge on tyres	Tax for remediation of contaminated sites
Manitoba – Non-deposit containers tax	Product charge on refrigerators and refrigerants	Incentive tax on volatile organic compounds
Manitoba – Tyres tax	Water pollution levy	
New Brunswick – Tyres tax		**UNITED KINGDOM**
Nova Scotia – Tyres tax	**ICELAND**	Aggregate levy
Ontario – Alcoholic beverage container tax	Hazardous waste fee	Landfill tax
Prince Edward Island – Tyres tax	Recycling charge	
		UNITED STATES (*Examples only*)
CZECH REPUBLIC	**IRELAND**	Alabama – State severance tax
Air pollution fee	Plastic bag levy	Alabama– Local severance taxes
Payments for production and import of ozone depleting chemicals		Arkansas – Severance tax
	ITALY	Arkansas – Waste tyre fee
DENMARK	Aircraft noise taxes	Federal – Ozone depletion tax
Duty on raw materials	Charge on air pollution	Indiana – Solid waste management fee
Duty on certain chlorinated solvents	Tax on plastic bags	New Jersey – Landfill closure and contingency tax
Duty on certain retail containers	Tax on waste disposal	New Jersey – Public community water system tax
Duty on CFC, HFC, PFAC and SF_6		New Jersey – Spill compensation and control tax
Duty on disposable tableware	**NETHERLANDS**	New Jersey – Litter control tax
Duty on electric bulbs and electric fuses	Levy on water pollution	Utah – Mining severance tax

Source: OECD/EEA database on instruments used for environmental policy. Fees and charges are not included.

used more specifically for environmental purposes, in particular for the operation of waste collection or recycling systems, for the clean-up of contaminated sites, etc.

2.2. The revenues raised though environmentally related taxes

On average, the revenues raised from environmentally related *taxes* represent some 2-2.5% of GDP, but as can be seen from Figure 2.1, there are large differences from country to country. In the Czech Republic, Denmark, Finland, the Netherlands, Norway and – especially – Turkey, the revenues were higher than 3% of GDP in 2003,[4] while in the United States, these revenues represented less than 1% of GDP.

Figure 2.1. **Revenues from environmentally related taxes in per cent of GDP**
1995, 1999 and 2003

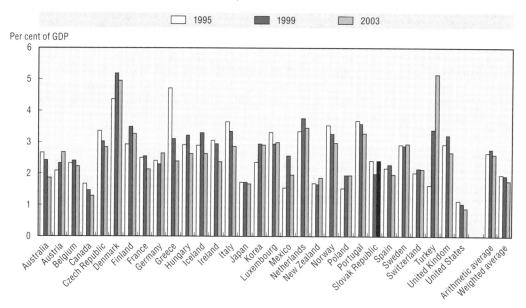

Source: OECD/EEA database on instruments for environmental policy. For the Slovak Republic, 2002 figures are used for 2003. The averages are calculated only across the countries for which 2003 numbers are available.

There is also a trend for the total revenues from environmentally related taxes on average to decrease slightly over time – even if a number of new taxes have been implemented since 1995, which in isolation should contribute to increase the revenues. The observed decline can in part be explained by a reduction in demand for petrol in OECD Europe. This is further discussed in Section 3.3.1 below.

It is emphasised that the amount of revenues raised is, at best, a very imprecise indicator of the "environmental friendliness" of the tax system in the country in question. From an environmental point of view one would generally like to see the size of the tax-base diminish through behavioural changes among firms and households – which would cause the revenues to decline. A country can, in principle, have low revenues from environmental taxes either because they apply low tax rates in the relevant taxes, or because they apply very high tax rates that have triggered large changes in behaviour. This being underlined, the countries with the lowest revenue shares in Figure 2.1 do not in general apply higher tax rates in their environmentally related taxes than other OECD countries do, *cf.* the discussion in Section 2.3.

Figure 2.2 shows the revenues from environmentally related taxes in per cent of total tax revenue. On average, this share is in the order of 6-7%, but again with very significant variations between countries. Turkey was in 2003, by far, the country that had the highest share of total tax revenue raised through environmentally related taxes – after a very strong growth in this share since 1995 – with Korea and Denmark following next. It is also worth noting that this share has decreased significantly since 1995 in Greece and Portugal – the two countries that raised the largest share of total tax revenue on environmentally related taxes in 1995.

Figure 2.2. **Revenues from environmentally related taxes in per cent of total tax revenue**

1995, 1999 and 2003

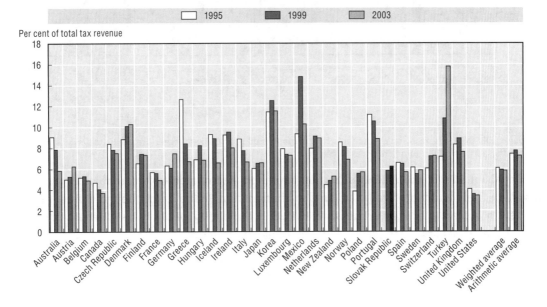

Source: OECD/EEA database on instruments for environmental policy. The averages are calculated only across the countries for which 2003 numbers are available. Information on total tax revenue is lacking for 1995 for the Slovak Republic, and 2002 figures are used for 2003 for this country.

In addition to the caveats mentioned above about making any judgements based on the amount of revenues raised, it is underlined that the share of revenues from environmentally related taxes in total tax revenue is – obviously – also influenced by the amount of revenue raised on other tax bases in a given country. If the revenues raised on other tax-bases are low, the share of total tax revenues raised through environmentally related taxes will be higher than otherwise.

Figure 2.3 illustrates the *nominal* amounts of revenue raised through environmentally related taxes per capita. That is, no correction is made for developments in the price level between 1995 and 2003. The large differences between countries does – of course – also reflect differences in income levels, in addition to varying weight placed on environmentally related taxation. However, also between countries at a relatively similar income levels are there large differences in the amount of revenues raised per capita.[5]

Figure 2.3. **Revenues from environmentally related taxes per capita**
1995, 1999 and 2003.

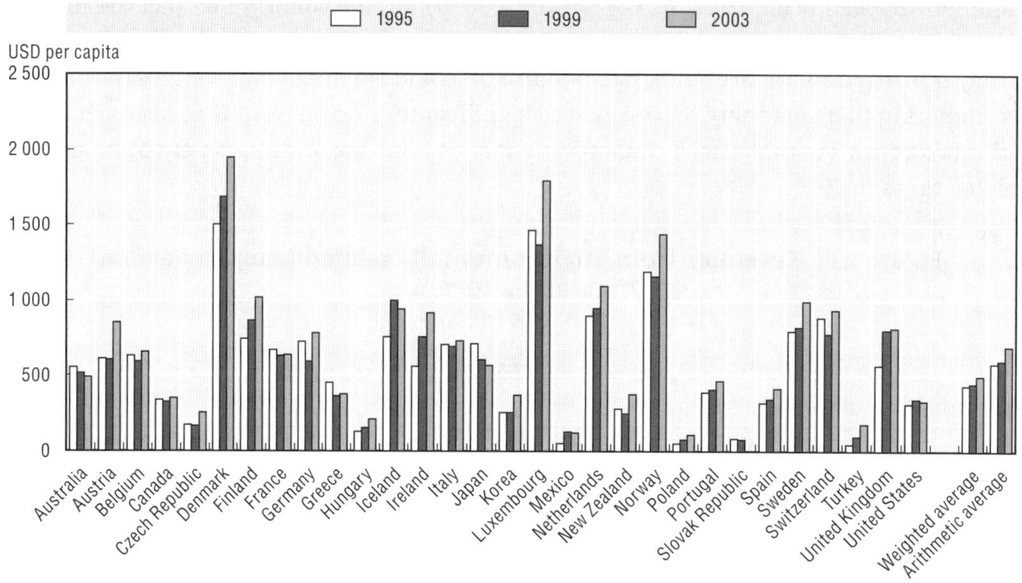

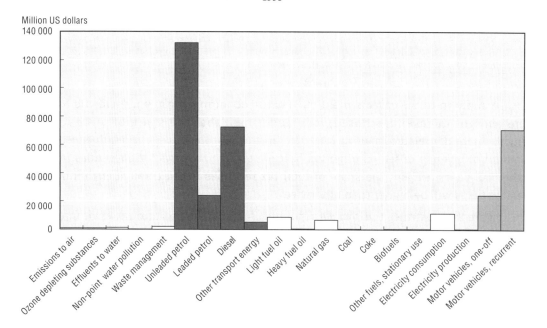

Source: OECD/EEA database on instruments for environmental policy. The averages are calculated only across the countries for which 2003 numbers are available.

About 90% of the revenues from environmentally related taxes are raised on motor vehicle fuels and motor vehicles. This is illustrated in Figure 2.4. Although the estimates used to prepare that graph date back to 1995, more aggregate analyses of recent revenue data make it clear that the picture largely remains the same.

Figure 2.4. **Revenues raised on different tax-bases**
1995

Source: Estimates made in OECD member countries. Although the estimates used to prepare that graph date back to 1995, analyses of recent revenue data make it clear that the picture largely remains the same.

A few changes have, however, taken place. Hardly any revenue is now raised on leaded petrol, as this tax-base has largely disappeared from the market in almost all OECD member countries. The leaded petrol has most often been replaced by unleaded petrol and by diesel, so the share of these tax-bases has certainly increased somewhat. Due to an on-going shift from petrol-driven to diesel-driven cars, the increase in the share of diesel in total revenues from environmentally related taxes has been particularly important.

On a smaller scale, the last 10 years have also witnessed a considerable growth in the revenues raised on waste-related tax-bases in some OECD member countries – i.e. either through taxes on the final treatment of waste (incineration and/or landfilling) or through product-taxes levied on certain products due to particular challenges they represent for waste management (batteries, tyres, lubricant oil,...). This is illustrated in Figure 2.5, which depicts the revenues raised through waste-related taxes in 9 selected OECD countries. Whereas revenues from waste-related taxes represented about 0.4% of all the revenue from environmentally related taxes in these countries in 1995, this share had increased to 2.9% in 2003. The amount of revenue raised through the waste-related taxes in these countries did in fact correspond to 0.7% of all the revenues raised through environmentally related taxes in *all* OECD member countries in 2003.

In 2003, about 45% of the revenue from waste-related taxes stemmed from taxes on individual products while about 55% was raised through taxes on incineration or landfilling.

Figure 2.5. **Revenues raised through waste related taxes**

In per cent of total revenues from environmentally related taxes, 1994-2003, Austria, Czech Republic, Denmark, Finland, Netherlands, Norway, Sweden, Switzerland and the United Kingdom

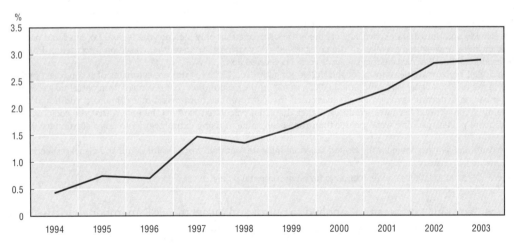

Source: OECD/EEA database on instruments for environmental policy.

2.3. **Some important environmentally related tax-bases**

2.3.1. *Motor fuels*

As already mentioned, a very significant share of all the revenues from environmentally related taxes arises from taxes on motor fuels. Such taxes were introduced in all member countries many decades ago, primarily as a means to raise revenue. Regardless of that, they do impact on the prices (potential) car users are facing, and thus they do have important environmental impacts. Figure 2.6 presents a comparison

Figure 2.6. **Tax rates on petrol and diesel in OECD member countries**
1.1.2000 and 1.1.2005, € per litre

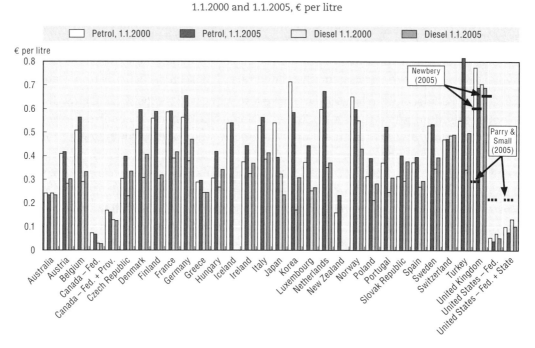

Notes:

Where several categories or classes of petrol or diesel exist, the graph shows the tax rate for the variety believed to be most environmentally friendly.

It is emphasised that developments in tax rates expressed in euro over time for countries outside the euro area can both be due to changes in tax rates in national currencies and to changes in the exchange rates.

There was no taxation of diesel fuel in Iceland and New Zealand as of 1.1.2005. Separate taxes were instead levied on the use of diesel-driven vehicles. No information on the tax rates applied is available for Mexico.

For Canada and United States, two sets of bars are shown; one that only includes the federal tax rates and one that also includes unweighted averages of the taxes levied at a provincial or state level, based on information from The International Fuel Tax Association, see *www.iftach.org/index50.htm*.

The dotted horizontal lines shown for the United Kingdom and United States are estimates of second-best optimal petrol tax rates made by Parry and Small (2005), made on the assumption that revenues from petrol taxes replaces revenues on distorting taxes on labour income. If instead revenues from petrol taxes financed additional public spending, the optimal tax rates would be higher than that calculated here (to the extent that the social value of additional public spending were greater than the social value of using extra revenue to cut distortionary income taxes).

The continuous lines shown for the United Kingdom indicate "optimal" tax rates for petrol and diesel respectively as estimated by Newbery (2005).

Source: OECD/EEA database on instruments for environmental policy.

of the "normal" tax rates that applied to petrol and diesel in OECD member countries as of 1.1.2000 and 1.1.2005.[6] Several comments can be made:

- The tax rates on motor fuels vary considerably between countries. For example, even when taking into account the taxes levied at a state or provincial level in Canada and USA, the taxes on petrol and diesel in these countries are only a small fraction of the taxes levied in several European countries – with the highest rates for petrol being found in Turkey and for diesel in the United Kingdom. Some comments on "optimal" fuel tax rates can be found in Box 2.2.

- There have also been significant changes in the tax rates between 1.1.2000 and 1.1.2005 in a number of countries – in both directions. Quite significant rate increases – measured in *national* currency – have *e.g.* taken place in the Czech Republic, Denmark, Germany, Hungary, New Zealand, Poland, Portugal and the Slovak Republic, while the tax rates for

Box 2.2. **"Optimal" tax rates on motor fuel**

Ideally, the tax rate on motor fuels should – as for other tax-bases – reflect the size of the negative environmental externalities associated with each fuel (and take account for the fact that any revenues raised can be used to lower other, more distorting, taxes, for example on labour incomes). It is, however, always difficult to estimate exactly what would be an "optimal" level for the different tax rates. Apart from data difficulties, there is the issue of the extent to which "optimal" tax rates should take account of concerns related to industrial competitiveness and income distribution. The studies referred to in this chapter do not take these issues into account, and so the term "optimal" should be interpreted accordingly. The level of optimal tax rates also will depend on the way the income from the taxes is used. For instance Parry & Small (2005) made the assumption that revenues from petrol taxes replace revenues from distorting taxes on labour income. If, instead, revenues from petrol taxes financed additional public spending, the optimal tax rates would be higher than that the rates they have calculated.

In the right-hand part of Figure 2.6, estimates of optimal petrol tax rates made by Parry and Small (2005) as regards the United Kingdom and United States are depicted. The congestion externality is the largest component in both nations according to these calculations, and the higher optimal tax for the United Kingdom is due mainly to a higher assumed value for marginal congestion cost. Revenue-raising needs also play a significant role, as do accident externalities and local air pollution. According to the findings of Parry and Small (2005), the tax rates on petrol are considerably below the "optimum" in the United States, while the petrol tax rate in the United Kingdom – according to this study – appears to be higher than "optimal".

Newbery (2005) includes a discussion of the estimates prepared by Parry and Small (2005). He states *inter alia* that the "pure road charge and green tax elements would amount to" EUR 0.6 per litre petrol and EUR 0.67 per litre diesel in the United Kingdom. He further states that if "this applied generally across the EU, The Netherlands and Germany would be taxing gasoline at about the right rate and only the UK is overcharging gasoline... All countries except the UK are probably undercharging diesel". The estimates prepared by Newbury (2005) do not include any so-called "Ramsey component" or the additional taxation justified by the impact of road user charges on labour supply.

EEA (2006) presents estimates of the externalities caused by different vehicle categories, expressed per vehicle-kilometre driven.

Recent work on optimal transport pricing place emphasis on variable road-user charges to *partly* replace taxes on fuels – *inter alia* in order to better reflect differences in the negative externalities over the day and depending on the location of the transport activity. See, for example, ECMT (2000, 2003), Glaister and Graham (2004), Newbery (2005) and Parry (2005). Several OECD member countries have recently put in place road-user charges for heavy goods vehicles (See for example OECD 2005a), and other countries prepare to do so.

Due to the larger negative environmental impacts of diesel use, the optimal tax rate on diesel could be significantly higher than the rates Parry & Small estimated for petrol.

The fact that diesel-driven vehicles are more energy-efficient than petrol-driven vehicles – and thus cause lower CO_2 emissions – is *not* an argument in favour of a lower tax rate for diesel than for petrol, as this effect is already *fully internalised* in the user-costs of the vehicles. That is, the users of a diesel-driven car benefit directly from the fact that a diesel motor is more fuel-efficient than a petrol motor, and no tax preference for diesel vs. petrol is thus required to provide the "right" incentive for the choice of fuel. In any case, according

> ### Box 2.2. **"Optimal" tax rates on motor fuel** (*cont.*)
>
> *e.g.* to Parry and & Small (2005), CO_2 emissions represent only a small share of the value of the negative environmental impacts related to the use of motor fuels.
>
> On the other hand, drivers obtain no private benefits from taking into account the environmental advantages of petrol vis-à-vis diesel – in relation to *e.g.* NO_x, particles and noise. Hence, a *higher* tax rate for diesel than for petrol would be required to provide a "correct" incentive structure from an environmental point of view. Ongoing efforts to reduce emissions of particles and NO_x from (in particular) diesel-driven vehicles will – of course – tend to reduce the "optimal" tax rate, if the vehicles perform as intended in real world driving conditions, which remains to be seen.

both petrol and diesel decreased substantially between 1.1.2000 and 1.1.2005 in Norway. The increases that have taken place in some of the new member States of the European Union come in response to the recent EU Directive on harmonisation of energy taxes and electricity, see CEU (2003).[7]

- The tax rate for diesel is much lower than the tax rate for petrol in most countries – with notable exceptions for Australia, Switzerland, the United Kingdom and United States. From an environmental point of view, this is regrettable, as diesel-driven vehicles cause more local air pollution and are noisier than petrol-driven vehicles.

The very significant raise in world crude oil prices and in the prices of petroleum products between 1998 and 2005 has triggered increased public opposition to the current taxes on motor fuels in many OECD member countries. Some comments in this regard are made in Box 2.3.

2.3.2. *Rate differentiation in motor fuel taxes*

A number of countries have also introduced rate differentiations in their petrol and/or diesel tax rates based on environmental criteria. For example, in Denmark the tax rates vary depending on the whether or not the petrol station has a system to capture the vapour from the petrol pumps.

A more common distinction is to vary the tax rate depending on the sulphur content of the fuels. As shown in Figure 2.7, 13 OECD member countries have done so. For diesel, some of the countries apply three steps in their taxes: One tax rate for diesel with a sulphur content above 50 mg per kg fuel (or 50 ppm), a lower tax rate for diesel where the sulphur content is between 10 and 50 mg per kg, and a still lower tax rate for diesel where the sulphur content is below 10 mg per kg (or 10 ppm). The large differences in the size of the "premium" given to lower-sulphur fuels between different countries is remarkable – especially as concerns diesel with a relative high sulphur content.

The fact that the tax differentiation is larger within a given country above and below 50 mg per kg than above and below 10 mg per kg is not surprising – as this "premium" *e.g.* trigger a larger absolute reduction in the sulphur content. If no rate differentiation is applied – and the sulphur content of the fuel is not limited through other instruments – there would often be a sulphur content of some 350 mg per kg in the diesel used in OECD member countries.

The impact of such tax rate differentiation can be quite spectacular – as illustrated in the case of the United Kingdom in Figure 2.8. After the introduction of a lower tax rate for

---

> ### Box 2.3. **Taxes on fuels when crude oil prices are high**
>
> Strong increases in world crude oil prices between 1998 and 2006 have led to strong public demand and political pressures in many OECD countries arguing for a decrease in current tax rates on motor fuels – as well as for other fuels. In this connexion, the claim is often made that it would be "fair" that governments reduce excise taxes on fuels, because higher net-of-tax fuel prices increase revenues from value added taxes and other *ad valorem* taxes. A number of comments can be made on this issue:
>
> - First of all, it is by no means given that total public revenue increase with increasing fuel prices. On the contrary, it is likely that the opposite will be the case. This is in part because any increase in VAT revenue collected on fuels will tend to be counter-balanced by a decrease in VAT revenue collected on other goods and services, as the share of payments for fuels in total household expenditures will tend to increase with increasing fuel prices. In addition, a fuel price increase will lead households and firms to reduce their total fuel use somewhat, which will cause revenues from fuel excise taxes to decline, even it the tax rates are kept unchanged.
>
> - The "need" for any reductions in fuel excise tax rates to compensate for alleged negative sectoral competitiveness impacts is questionable. The recent increases in crude oil and fuel prices are global phenomena. Hence, the foreign competitors of the sectors said to be "at risk" will generally face similar input price increases, and in well-functioning markets the increased costs of inputs will eventually to a large extent be passed on to the prices of products and services produced. Rather than granting additional tax reductions to sectors that in many cases already are given a preferential treatment – and where the negative environmental externalities related to their activity are largely *not* "internalised" in the prices – it would seem better to scale back or remove any provisions that prevent the relevant markets from working properly.
>
> - If oil importers start to reduce taxes in order to stabilise tax-inclusive fuel prices, oil exporters will know that they can at no risk increase their resource rents by restraining their production and thus increase crude prices further. Normally such actions would trigger reductions in demand that could reduce the incomes of the oil producers, but the demand reductions will be absent if tax-inclusive user prices are kept stable by tax reductions.
>
> - Strong increases in fuel prices can, however, lead to increased hardship for poorer households. Rather than addressing these concerns by (general) reductions in fuel taxes, it would seem better to apply instruments targeted directly at the households one wishes to reach. Income distribution impacts of environmentally related taxes are discussed further in Chapter 7 below.
>
> In September 2005, finance ministers of the European Union member states agreed not to lower fuel taxes in response to the higher crude oil prices.

low-sulphur diesel and petrol in 1999 and 2001 respectively, the higher-sulphur varieties rapidly disappeared from the market.[8] Similar developments have been observed in other countries.

2.3.3. Motor vehicles

As can be seen in Figure 2.4, motor vehicles constitute another major tax-base of environmentally related taxes. As for motor fuels, taxes on motor vehicles were often first introduced many years ago, primarily in order to raise revenues. Nevertheless, such taxes

Figure 2.7. **Differentiation in tax rates for petrol and diesel according to the sulphur content**

1st January 2005

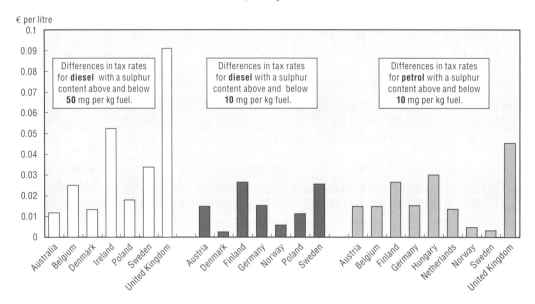

Source: OECD/EEA database on instruments used for environmental policy.

Figure 2.8. **Tax revenues raised on different motor fuels in the United Kingdom**

1994-2003, million GBP

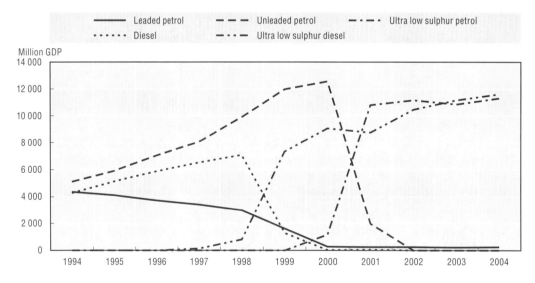

Source: www.statistics.gov.uk/StatBase/Expodata/Spreadsheets/D5688.xls.

do have significant environmental impacts. There are two main categories of taxes levied on motor vehicles: They can either be *one-off* or *recurrent*. One-off taxes are typically levied at the first sale or at the first registration of a motor vehicle in the country concerned. Recurrent taxes can be imposed at every later sale or registration of the vehicles, or they can be periodical, like annual taxes the owners have to pay to be allowed to use a vehicle during a given year.

For OECD as a whole, the amount of revenue collected through recurrent taxes on motor vehicles is much larger than the amount collected on one-off taxes, but in Denmark, Finland, Norway, Ireland, Korea and Portugal the opposite is the case.

The tax rates applied to first-time motor vehicle purchases are particularly high in Denmark and Norway. In Denmark the tax rate levied on the first registration of a large passenger car was EUR 8 683 + 178% of the tax value of the car exceeding EUR 8 500 as of 1.1.2005. The tax on first-time registration of passenger vehicles in Norway is outlined in Box 2.4.

As is the case in Norway (at least to some extent) – and opposed to what is the case in *e.g.* Denmark – the tax rates in a tax on motor vehicle purchases can vary according to environmental criteria. The weight, cylinder volume and motor power part of the tax all give the potential car purchasers an incentive to choose a car that is likely to use less fuel – and thus *i.a.* cause less greenhouse gas emissions.[9]

Box 2.4. The tax on the first registration of passenger vehicles in Norway

The structure of the tax on the first registration of motor vehicles in Norway is rather complicated. This box outlines the rates that apply to passenger vehicles. For each vehicle there is one part of the tax related to the cylinder volume of the engine, one part depending on the kW power of the motor and one part levied according to the weight of the vehicle. As of 1.1.2006, for a hypothetical passenger car weighing 1 600 kg with a cylinder volume of 2 000 cm^3 of a motor producing 140 kW the tax rate would be

- EUR 9 051 + EUR 20.54 per kg above 1 500 kg, *i.e.* EUR 11 105 according to the weight part;
- *plus* EUR 3 612 + EUR 8.03 per cm^3 above 1 800 cm^3, *i.e.* EUR 5 218 according to the cylinder volume part;
- *plus* EUR 9 196 + EUR 210.55 per kW above 130 kW, *i.e.* EUR 11 301 according to the motor power part.

Hence, the total tax rate for the first-time registration of such a vehicle would be EUR 27 624.

In Austria, a more direct link between the tax rate and the fuel consumption is included in the tax on motor vehicle registration – at least for passenger cars with fuel consumption of 10 litres or less per 100 km for diesel-driven vehicles and 11 litres or less per 100 km for petrol-driven vehicles. For these passenger vehicles, the fuel consumption per 100 km enters explicitly into the formula used to calculate the tax rate – which also is multiplied by 2% of the net purchase price. For passenger vehicles with higher fuel consumption than the limits indicated above, the tax rate is simply 16% of the net purchase price.

Also the tax rates in recurrent motor vehicle taxes can vary according to criteria of environmental relevance, such as the fuel consumption of the vehicle, the type of fuel used (diesel, petrol, LPG, etc.), the weight of the vehicle, whether or not the vehicle has a catalytic converter, etc. For example, in Austria the tax rate vary according to the kW power of the engine of the vehicle. In Denmark, Sweden and in some cantons in Switzerland, the annual tax on motor vehicle use varies with the weight of the vehicle. In Germany, the tax rates in the annual motor vehicle tax depend on the cylinder volume of the passenger vehicles, the so-called "Euro classification" of the vehicle and on the fuel the vehicle uses. Iceland and New Zealand levy recurrent taxes that depend on the number of km driven by diesel-driven vehicles – instead of any taxes on diesel fuel.[10]

In CEC (2005), a proposal is made for a new EU Directive with the double purpose of improving the functioning of the Internal Market and to implement the Community's strategy to reduce CO_2 emissions from passenger cars. The proposal does not intend to introduce any new passenger car related taxes, but to restructure such taxes if they are applied by member States, without obliging them to introduce such taxes. The European Commission proposes that by 1 December 2008 (the start of the Kyoto period) at least 25% of the total tax revenue from passenger car registration and annual circulation taxes respectively should originate in the CO_2 based element of each of these taxes. By 31 December 2010, at least 50% of the total tax revenue from both the annual circulation tax and the Registration tax (pending its abolition) should originate in the CO_2 based element of each of these taxes – according to the proposal.

In addition to motor vehicle taxes, several countries have recently introduced fees or charges on heavy goods vehicles that depend on the number of km driven. This is, for example the case in Austria, Germany and Switzerland.[11] The draft EU Directive 2003/175/COD[12] also encourages the application of road charges for heavy goods vehicles on all motorways and main national roads.

In the United Kingdom, London and Durham have introduced road toll systems that place a significant fee on any vehicle that enters the city centres. Somewhat similar schemes – but primarily designed to raise revenue rather than to affect driving patterns – have been in use a number of years in the largest towns in Norway. Congestion charges trials started in Stockholm, Sweden, on 3 January 2006 and the trials will be concluded by 31 July 2006.

2.3.4. Waste

As mentioned above, the use of waste-related taxes is increasing in a number of OECD member countries. Such taxes can be levied on certain products that cause particular problems for waste management, in order to discourage their use.[13] They can also be levied on the final treatment of waste – on top of "tipping fees" that landfill operators charge to cover their private costs – in order to "internalise" the negative environmental impacts related to incineration and/or landfilling.[14]

Figure 2.9 illustrates developments in tax rates on landfilling of municipal or household waste in some member countries. It is clear that the use of landfill taxes is getting more widespread, and that the tax rates in many cases have been increased over the years. Box 2.5 provides some comments on the "optimal" level of landfill taxes.

Like a few other countries, Norway has also introduced taxation of waste incineration – in addition to the tax on landfilling. An interesting and useful new development is that the tax on incineration is levied according to measured or estimated emissions of a number of pollutants coming out of the incinerators rather than on the amount of waste delivered for incineration. This gives the operators of the incinerators an incentive to lower the emissions per tonne of waste burned.[15]

Figure 2.9. **Tax rates on landfilling of municipal waste**

EUR per tonne, 2000, 2002 and 2005

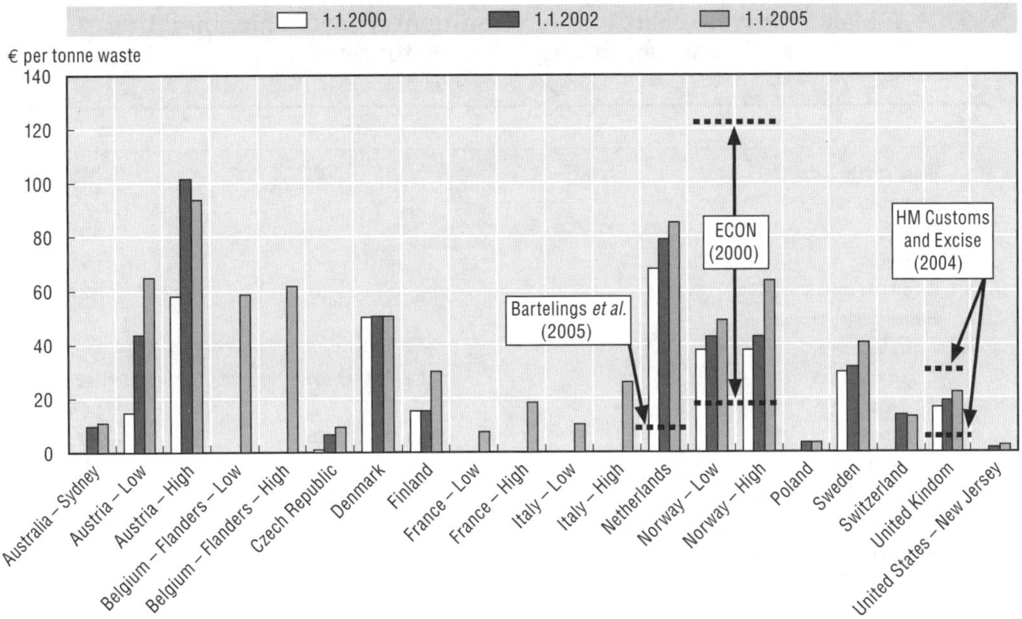

Note: Several countries differentiate their tax rates depending on the quality of the environmental quality of the landfill site. This is represented in the figure with a low tax rate that apply to the deposits with the highest environmental quality, and a high tax rate, for deposits with the lowest standards that are authorised to receive household waste.

Source: OECD/EEA database on instruments for environmental policy.

Box 2.5. **"Optimal" tax rates on the landfilling of household waste**

Similarly to the case of motor fuels discussed in Box 2.2, the "optimal" tax rate on the landfilling of waste would reflect the economic value of the negative environmental impacts caused by landfilling. The quantification of these impacts is obviously not simple, but a few estimates are available, as illustrated by the dotted horizontal lines included in Figure 2.9.

HM Customs & Excise (2004) presents a low and a high estimate for the externalities relating to landfilling in the United Kingdom, building on a scientific study of the health and environment impacts of different waste management options and a survey of recent valuation studies. In summary, it was found that "in the central case, the external costs of landfill may be around GBP 10 per tonne of municipal solid waste and the external costs of incineration (with energy recovery) may be around GBP 13 to GBP 14 per tonne of municipal waste".

It was further said that "sensitivity analysis around the central case shows that the range of possible externality estimates is large. For example, varying the emissions levels from landfill sites and incinerators, and holding all other factors constant, gives a range of external costs of GBP 5 to GBP 20 (EUR 7.4 to EUR 29.5) per tonne for landfill and GBP 7 to GBP 21 (EUR 10.3 to EUR 30.9) per tonne for incineration".

Dijkgraaf and Vollebergh (2004) and Bartelings *et al.* (2005) have compared the private and social costs of landfilling and incineration in the Netherlands, see the table below. These estimates concern a *new site*, using *state-of-the-art technologies*, in accordance with current Dutch emission standards, etc.

Box 2.5. "Optimal" tax rates on the landfilling of household waste (cont.)

Table 2.2. Private costs and environmental costs of incineration and landfilling in the Netherlands

Euro per tonne waste

		Landfilling	Incineration
1.	**Gross private costs**	**40**	**125**
2.	Private cost savings		
3.	– Energy	–4	–21
4.	– Materials	–0	–3
5.	**Net private costs** [1 – (3 + 4)]	**36**	**101**
6.	**Environmental impacts**		
7.	– Climate change	4.21 *(1.46 – 54.50)*	0.11 *(0.06 – 0.88)*
8.	– Other emissions to air	1.22 *(0.58 – 1.85)*	7.22 *(1.50 – 7.22)*
9.	– Transport-related impacts	1.25	1.67
10.	– Disamenity	3.50 *(3.50 – 3.80)*	9.09 *(9.09 – 9.87)*
11.	– Solid waste		0.11 *(0.09 – 5.62)*
12.	– Land use	0.00 *(0.00 – 17.88)*	
13.	**Gross environmental costs** (7 + 8 + 9 + 10 + 11 + 12)	**10.18** *(6.79 – 79.28)*	**18.20** *(12.41 – 25.26)*
14.	**Environmental cost savings related to energy generation and materials recovery**	**–1.14** *(–0.85 – –4.46)*	**–7.63** *(–6.96 – –12.56)*
15.	**Net environmental costs** (13 – 14)	**9.04** *(5.94 – 74.82)*	**10.57** *(5.45 – 12.70)*
16.	**Net social costs** (5 + 15)	**45.04** *(41.94 – 110.82)*	**111.57** *(106.45 – 113.7)*

Source: Based on Bartelings *et al.* (2005).

The gross private costs of building a new incinerator were in both studies found to be much higher per tonne of waste than for building a new landfill. The difference in private costs is reduced when taking into account the economic value of the energy produced by the different facilities, and of the materials recovered from the ash of the incinerator – but only to a modest extent. Hence, the net private cost for a landfill was found to be EUR 36 per tonne waste, versus EUR 101 per tonne for an incinerator.

The numbers given as regards environmental costs are based on the estimates of Bartelings *et al.* (2005). Their "best" estimates are presented in the upper lines, while the ranges between their "low" and "high" estimates are given in the parentheses below.

The requirements on landfills in the Netherlands are so strict that no pollution of water was expected to take place from a new landfill. The "best" estimate regarding climate change impacts of landfilling is based on an assumed damage cost of EUR 10 per tonne CO_2, a CH_4/CO_2 damage ratio of 21, a CH_4 recovery rate of 42.5% and a 4% discount rate. The high-end estimate is based on an assumed damage cost of EUR 80 per tonne CO_2, a CH_4/CO_2 damage ratio of 30, a CH_4 recovery rate of 40% and a 3% discount rate.

The "best" estimates indicate that also the gross environmental externalities of incineration are significantly higher than what is the case for landfilling – largely due to higher emissions to air other greenhouse gases, and because the disamenities of the incinerators are found to affect a larger number of people.

However, incineration also causes important environmental savings in relation to the energy and materials that are recovered in the process. Hence, the net environmental costs of landfilling are only slightly lower than what they are for incineration – EUR 9 *vs.* EUR 10.5 per tonne waste. The estimated environmental costs of landfilling are, however, only a small fraction of the current tax rate of the Dutch Landfill Tax.

> **Box 2.5. "Optimal" tax rates on the landfilling of household waste** (cont.)
>
> Adding together the net private costs and the net environmental costs provides the net social cost, here estimated to be approximately EUR 45 per tonne for landfilling and EUR 112 per tonne for incineration.
>
> One can notice that the estimates of the value of the environmental externalities of both landfilling and incineration presented in the Dutch studies are well within the range presented in HM Customs and Excise (2004). The differences in *environmental costs* of landfilling and incineration are in all cases so small that the ranking of the two options in terms of their *net social costs* is likely to depend mostly on the *private costs* of landfilling *vs.* incineration.
>
> For a further discussion of waste policy instruments, see OECD (2004a and 2004b) and OECD (Forthcoming).

The dotted horizontal lines in Figure 2.9 illustrate various recent estimates of the environmental externalities related to the landfilling of household (or municipal) waste. Bartelings *et al.* (2005) estimated these externalities related to a new landfill in the Netherlands. ECON (2000) presents estimates for both existing and new deposits in Norway – with the (very) high estimate related to old-fashioned existing sites. HM Customs & Excise (2004) presents a low and a high estimate for the externalities relating to landfilling in the United Kingdom, building on DEFRA (2004a, 2004b and 2004c), a scientific study of the health and environment impacts of different waste management options and a survey of recent valuation studies.

The taxes on incineration and landfilling are at first hand paid by the operators in the waste management system (site owners, municipalities, companies that deliver their waste for final treatment, etc.). As concerns household waste and other municipal waste, such taxes can give municipalities an incentive to put in place separate collection schemes and other measures to promote waste prevention and recycling, thus reducing the amount of waste that has to be sent for final disposal.[16] Unless coupled with waste collection charges that vary with the amount of waste for final disposal, such taxes do, however, not give households a direct incentive to reduce the amount of waste they generate. The collection charges ought also somehow to be differentiated according to the environmental damage caused by various waste fractions. They ought to be lower for fractions that are not delivered for final disposal, but for instance recycled (paper, etc.), than for residual wastes that has to be landfilled or incinerated. Variable waste collection charges are becoming more widespread in countries, but in for instance the United Kingdom, they are not allowed.[17]

If improperly managed, many types of batteries can cause considerable environmental harm at the waste stage, and a number of taxes and fees/charges are thus levied on small and large batteries or accumulators. Table 2.3 gives an overview over of the relevant tax-bases that are used in a number of OECD countries, and the tax-rates that apply to them. Due to the large number of very different tax-bases and to the application of tax rates both per unit and according to weight, it is not easy to make direct comparisons. It is, however, clear that there are significant variations in the tax rates applied between different countries. In particular, the rates applied to some types of batteries in Sweden are high compared to what is found in other countries.

Table 2.3. **Taxes, fees and charges levied on batteries in OECD member countries**

Austria – Charge on batteries	
Button cell batteries, up to 5 g	0.01 EUR per unit
Consumer batteries below 25 g	0.02 EUR per unit
Consumer batteries between 25 and 100 g	0.07 EUR per unit
Consumer batteries between 100 and 450 g	0.36 EUR per unit
Video-packs and battery combinations of more than 500 g	0.73 EUR per unit
Belgium – Environmental taxes	
Batteries	0.5 EUR per unit
Canada – British Columbia – Batteries tax	
Vehicle batteries	3.09 EUR per unit
Denmark – Charge on batteries	
Lead batteries – car batteries < 100 Ah	1.61 EUR per unit
Lead batteries – car batteries > 100 Ah	3.23 EUR per unit
Lead batteries – other	2.42 EUR per unit
Nickel-cadmium round cells	0.81 EUR per cell
Hungary – Product charge on batteries	
Batteries used in vehicles	0.15 EUR per kg
Iceland – Hazardous waste fee	
Alkaline button batteries	0.10 EUR per unit
Batteries containing acid	0.24 EUR per kg
Batteries containing mercury or nickel cadmium	2.30 EUR per kg
Batteries in instruments	1.33-15.47 EUR per unit
Batteries in voltage transformers	0.24 EUR per kg
Korea – Waste disposal charge	
Lithium and nickel batteries, up to 20 g	0.01 EUR per unit
Lithium and nickel batteries, over 20 g	0.60 EUR per kg
Mercury batteries	0.08 EUR per unit
Silver oxide batteries	0.05 EUR per unit
Poland – Product charges	
Galvanic cells and batteries	0.02-1.13 EUR per unit
Nickel-cadmium accumulators	0.07-5.88 EUR per unit
Portugal – Charge on batteries	
Button batteries	2.56 EUR per kg
Lithium batteries	1.40 EUR per kg
Lithium ions batteries	0.70 EUR per kg
Nickel-cadmium + NiMH batteries	0.79 EUR per kg
Saline + Alkaline batteries	1.12 EUR per kg
Slovak Republic – Product charge for recycling and waste management	
Batteries and accumulators	0.20 EUR per kg
Sweden – Battery fee	
Environmentally harmful batteries – starter batteries that contain lead	3.29 EUR per kg
Environmentally harmful batteries – other lead batteries	0.19 EUR per kg
Environmentally harmful batteries – sealed nickel cadmium batteries	32.9 EUR per kg
Environmentally harmful batteries – silver oxide batteries	54.8 EUR per kg
Environmentally harmful batteries – zinc air batteries	54.8 EUR per kg
Switzerland – Prepaid fee on batteries	
Batteries	2.07 EUR per kg

Source: OECD/EEA database on instruments used for environmental policy.

2.4. Tax exemptions, refund mechanisms, rate reductions, etc.

To get a correct picture of which "effective" tax rates various categories of firms and households are facing it is also important to take into account the many exemptions, refund mechanisms, etc., that are included in many of the environmentally related taxes. The OECD/EEA database includes more than 1 150 exemptions in the 375 or so taxes applied in OECD countries.[18] This information must be interpreted with caution: exemptions are introduced for a number of social, environmental and economic reasons – including a concern for the international competitiveness of certain sectors – and it can often be difficult to pin-point which are the most important motivations.

Much of the information in the OECD/EEA database has been linked to one or more ISIC production sectors.[19] In all more than 1 800 links have been made between a given sector and an exemption – either because the sector *produces* a product that is exempted, or because it is a *major user* of an exempted product. Only for about 50 exemptions has it not been found meaningful to make a link to one or more ISIC sectors – but many of the exemptions to which one or more sectors have been linked, can have been introduced primarily for other concerns than the international competitiveness of these sectors.

Table 2.4 details the number of exemptions that a given sector has been linked to, and it also distinguishes three main types of taxes in which an exemption is given. The largest number of exemptions is linked to the sector "Manufacture of coke, refined petroleum products and nuclear fuel". This is due to the large number of exemptions for various fossil fuels, which are produced by this sector. The high number of exemptions linked to the sector "Manufacture of motor vehicles, trailers and semi-trailers" is due to the many exemptions that exist in various motor vehicle taxes. The significant number of exemptions in fuel taxes linked to sea and air transport is, on the other hand, due to the use these sectors make of exempted products.

The OECD/EEA database also records about 175 refund mechanisms in the environmentally related taxes in all. A number of these refunds have been introduced for competitiveness reasons, others for social reasons, etc. The refunds have been linked to various ISIC sectors in a similar way to the exemptions described above. All in all, about a 150 links have been made between sectors and refunds, and these are also detailed in Table 2.4. Here again, it is difficult to ascertain whether a given refund is specifically crafted to alleviate possible competitiveness effects: while a few refunds aim at "rewarding" environment friendly practices or processes (*e.g.*, taxes paid on LPG, natural gas, low-sulphur and sulphur-free diesel and electricity used in public transportation are reimbursed in Denmark), many other refunds are designed to lighten the tax burden of industry under specific conditions.

Rather than full exemptions or 100% tax refunds, many countries grant significant reductions in the tax rates that apply to certain sectors or uses. For example, a "dyed" or coloured type of diesel is sold at low tax rates for off-road uses, *e.g.* in agriculture, in several countries. The tax rate on fuel used for *domestic* aviation – if such fuels are taxed at all[20] – are generally much lower than the tax rates for fuels used in other sectors. Energy tax rates facing energy-intensive sectors – if they exist at all – are also generally significantly lower than those facing other actors. For instance, in Germany, energy-intensive firms only pay 3% of the "normal" tax rates of the energy taxes introduced as part of the ecological tax reform, while other manufacturing firms have been granted a 60% rate reduction.

Tax rate reductions are sometimes coupled with a condition that the firms in question have to improve their environmental performance through other means. For example, energy-intensive firms in the United Kingdom can benefit from an 80% rate reduction in the Climate Change Levy if they reach targets set in negotiated Climate Change Agreements.[21] In Denmark, registered businesses can obtain a reimbursement of 13/18 of the duty on CO_2 as regards products used in energy-intensive processing, and additionally

Table 2.4. **Exemptions in environmentally related taxes by sector
and type of tax-bases affected**

Sector name	Exemptions in all	Exemptions in:			Refunds in all
		Fossil fuel taxes	Motor vehicle taxes	Waste taxes	
Agriculture, hunting and forestry	78	42	17	4	20
Fishing	36	26	4	1	4
Mining of coal and lignite; extraction of peat	35	19	1	2	0
Extraction of crude petroleum and natural gas, and incidental service, ex. surveying	39	33	1	0	0
Mining of uranium and thorium ores	11	6	1	1	0
Mining of metal ores	11	7	0	1	0
Other mining and quarrying	20	8	1	5	0
All, or most, of manufacturing	49	34	2	1	23
Manufacture of food products, beverages and tobacco products	15	1	0	14	1
Manufacture of textiles, wearing apparel and footwear, etc.	0	0	0	0	0
Manufacture of wood and wood products, etc., except furniture	8	7	0	0	0
Manufacture of paper and paper products	14	4	0	9	2
Publishing, printing and reproduction of recorded media	2	0	0	2	0
Manufacture of coke, refined petroleum products and nuclear fuel	345	303	6	11	15
Manufacture of chemicals and chemical products	37	20	0	3	6
Manufacture of rubber and plastics products	15	0	0	15	2
Manufacture of other non-metallic mineral products	5	2	0	0	0
Manufacture of basic metals	10	6	0	1	0
Manufacture of fabricated metal products, except machinery and equipment	8	1	0	6	0
Manufacture of machinery and equipment n.e.c.	16	2	13	1	1
Manufacture of office, accounting and computing machinery	1	1	0	0	0
Manufacture of electrical machinery and apparatus n.e.c.	7	0	0	7	2
Manufacture of radio, television and communication equipment and apparatus	0	0	0	0	0
Manufacture of medical, precision and optical instruments, watches and clocks	1	0	0	1	0
Manufacture of motor vehicles, trailers and semi-trailers	211	2	208	0	6
Manufacture of other transport equipment	28	5	17	0	0
Manufacture of furniture; manufacturing n.e.c.	1	0	0	0	0
Recycling	16	0	0	10	0
Production, transmission and distribution of electricity	123	60	1	0	10
Manufacture of gas; distribution of gaseous fuels through mains	9	8	0	0	0
Steam and hot water supply	23	18	0	0	3
Collection, purification and distribution of water	5	0	1	0	0
Construction	19	1	9	4	0
Wholesale and retail trade; repair of motor vehicles, etc.	8	3	2	2	0
Hotels and restaurants	0	0	0	0	0
Transport via railways	26	17	6	0	4

Source: The OECD/EEA database on instruments used for environmental policy.

Table 2.4. **Exemptions in environmentally related taxes by sector and type of tax-bases affected** (cont.)

Sector name	Exemptions in all	Exemptions in:			Refunds in all
		Fossil fuel taxes	Motor vehicle taxes	Waste taxes	
Other land transport	78	22	50	2	23
Transport via pipelines	0	0	0	0	0
Sea and coastal water transport	49	40	4	1	3
Inland water transport	33	25	2	1	2
Air transport	71	42	8	1	4
Supporting and auxiliary transport activities; activities of travel agencies	5	3	1	1	0
Post and courier activities	0	0	0	0	0
Telecommunications	2	0	0	0	0
Financial intermediation	4	0	4	0	0
Real estate, renting and business activities	14	6	6	0	4
Public administration and defence; compulsory social security	175	50	90	7	4
Education	16	6	4	1	3
Health and social work	67	7	52	2	1
Sewage and refuse disposal, sanitation and similar activities	49	4	4	33	5
Other	18	4	9	0	0
Total	**1 813**	**845**	**524**	**150**	**148**

Source: The OECD/EEA database on instruments used for environmental policy.

a reimbursement of 11/45 of the tax if an energy-savings agreement is made with the Ministry of Transportation and Energy.[22] The use of environmentally related taxes in combination with other policy instruments is discussed further in Chapter 10.

A fourth type of mechanism that can impact significantly on the effective tax rates paid especially by industries is "tax ceilings" introduced in certain taxes in some countries. For example in Sweden, if a manufacturing company or an agriculture, forestry, fishing or greenhouse cultivator have a remaining CO_2 tax – after a 79% rate reduction grated to all firms in these sectors – that exceeds 0.8% of the turnover, the company then only has to pay 24% of the tax normally due above this ceiling.

Hohlhaas and Bach (2005) discuss the impacts of special provisions in the German ecological tax reform on the environmental effectiveness of the reform. Prior to 2003, goods-producing industries in general paid only 20% of the tax rates of other taxpayers. In addition, an energy-intensive firm would get a *complete exemption* for any energy tax payment that exceeded 1.2 times the reduction the firm had obtained in social security payments through the reform. Since 2003, the goods-producing firms pay 60% of the ordinary tax rates. In addition, energy-intensive firms now get a *95% reduction* in tax payments that exceed (1.0 times) the social security reduction they obtained. Hence, even the most energy-intensive firms now receive *some – albeit very modest* – incentive from the reform to reduce their energy use. The impacts of the modification are illustrated in Figure 2.10. While the net tax burden increased for some firms, for companies with an energy use between the points E_3 and E_5 in the graph, the net tax burden was lowered.

Eurostat (2003) presents an analysis of how much revenue from energy taxes was raised in different economic sectors in Denmark, Finland, Norway and Sweden. The main finding is shown in Figure 2.11. In all the four countries, the manufacturing sector accounts for a much smaller share of energy tax payments than their share in energy use. The opposite is the case for households, whereas the service sectors pay energy taxes more in proportion to their energy use.

Figure 2.10. **Special provisions for energy-intensive firms in the ecological tax reform in Germany**

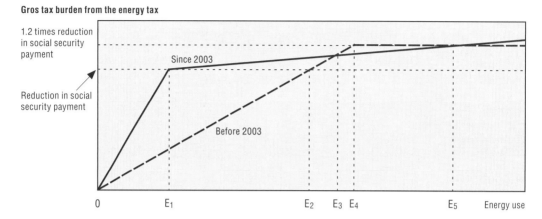

Source: Based on Kohlhaas and Bach (2005).

Figure 2.11. **Energy use and energy taxes paid**
Sweden, Norway, Finland and Denmark, 1999

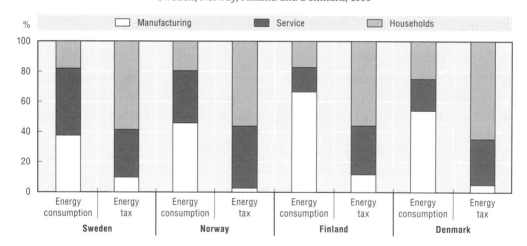

Source: Eurostat (2003).

Notes

1. It also covers more than 60 taxes and more than 180 fees and charges levied in some Central and Eastern European countries that are not OECD member countries. The database further provides information on almost 50 tradable permits systems (all in OECD countries), more than 50 deposit-refund systems (about 40 in OECD countries), more than 210 environmentally motivated subsidy schemes (all but 20 in OECD countries) and about 100 voluntary approaches (almost all in OECD countries).

   The database aims to cover instruments levied at both a national and a sub-national level (states, provinces, länder, etc.). The coverage of taxes and charges levied at a sub-national level has improved considerably in 2005 and 2006, but it is still not complete. The database is freely available at *www.oecd.org/env/policies/database*.

2. For a recent overview over economic instruments for environmental policy – applied also in a number of non-OECD member countries – see EEA (2005). That publication also reviews *inter alia* emission trading systems, subsidies and liability and compensation systems applied in European countries.

3. In addition, among the 250 environmentally related fees and charges in the database, some 75 are related to waste. The distinction between taxes and fees/charges can sometimes be difficult in the waste area, depending *e.g.* on whether the use of the revenues raised to establish a collection system for certain products is seen as a service provided to those that are asked to pay the levy in question.

4. The small decline indicated for Denmark between 1999 and 2003 was reversed in 2004, when the revenues from environmentally related taxes were equal to 5.1% of GDP.

5. The high amounts per capita found for Luxembourg do to a significant extent reflect purchases of motor vehicle fuels made by persons living in neighbouring countries. For example, while the tax rate for unleaded petrol was EUR 0.44 per litre in Luxembourg 1.1.2004, it was EUR 0.65 per litre at the same date in Germany, EUR 0.66 in the Netherlands, EUR 0.59 in France and EUR 0.51 in Belgium.

6. In many OECD countries, certain sectors pay significantly lower effective tax rates, in particular as regards diesel. This applies *e.g.* to the goods transport sector, public transport and to off-highway uses of vehicles, for instance in the agriculture sector.

7. Kouvaritakis *et al.* (2005) analyse impacts of energy taxation in the enlarged European Union. Kohlhaas *et al.* (2004) study effects of the EU Directive on energy tax harmonisation.

8. This obviously represented a *supply-side* response much more than a change in the demand from the individual vehicle users. In contrast to what earlier was the case regarding leaded and unleaded petrol, petrol stations generally do not offer customers a choice between high-sulphur and low-sulphur fuels – *inter alia* for logistics reasons.

9. CEC (2002a) discusses possibilities to introduce fiscal measures to reduce CO_2 emissions from new passenger cars.

10. From 1 July 2005 the tax system in Iceland was changed. The tax per km driven for diesel vehicles was discontinued and a tax of EUR 0.52 per litre diesel was introduced.

11. OECD (2005a) discusses how the obstacles to the introduction of the Swiss fee for heavy goods vehicles were overcome.

12. See *http://europa.eu.int/eur-lex/pri/en/oj/dat/2004/c_241/c_24120040928en00580064.pdf*. The European Commission issued a communication on this issue 7 September 2005, *cf.* COM(2005)423 final, available at *http://europa.eu.int/eur-lex/lex/LexUriServ/site/en/com/2005/com2005_0423en01.pdf*.

13. Sometimes the revenues raised through such levies are used to finance separate collection schemes for such products. Such developments can make it more difficult to draw a clear distinction between taxes and fees/charges (see Box 2.1), as – due to the collection-service provided – the levy might be classified as "requited", thus being a fee/charge.

14. Davies and Doble (2004) and Martinsen and Vassnes (2004) discuss the landfill tax in the UK and the tax on final treatment of waste in Norway respectively. Dijkgraaf (2004) and Dijkgraaf and Vollebergh (2004) discuss the landfill tax in the Netherlands.

15. One can, however, ask why not similar taxes are levied on the same type of emissions stemming from other (large) stationary sources. See Martinsen and Vassnes (2004) for further details on the incineration tax.

16. This assumes that municipalities have an interest in reducing the taxes paid by their citizens.

17. For a discussion of the impacts of variable waste collection charges, see OECD (2006a). For a further discussion of the links between taxes on final disposal and variable waste collection charges, see OECD (forthcoming).

18. The database also details more than 200 exemptions in the 250 fees and charges included from OECD countries, plus more than 100 and about 50 exemptions respectively in the 60 taxes and the 180 fees and charges from non-OECD member countries. The fact that there are fewer exemptions per fee or charge on average than for per tax could in part be because the information on fees and charges in general is less complete than what it is for taxes.

19. See *http://unstats.un.org/unsd/cr/registry/regcst.asp?Cl=17* for further information on ISIC.

20. Fuels used in *international* aviation are as of 2005 *not* taxed in any country.

21. It is worth noting that the Climate Change Levy does not apply to households at all. For a discussion of the political economy of this levy, see OECD (2005b) and Green Budget Germany (2004).

22. For a discussion of an earlier version of these energy savings agreements, see OECD (2003a).

ISBN 92-64-02552-9
The Political Economy of Environmentally Related Taxes
© OECD 2006

Chapter 3

Environmental Effectiveness

While the theoretical advantages (especially the static and dynamic efficiency) of environmental taxes are well known, *ex post* data on environmental effectiveness are still relatively scarce – although price elasticity estimates are available for many energy products. There are several reasons for this. In the first place, experience is often too recent to allow for a meaningful evaluation. Secondly, there is a shortage of data and practice when it comes to policy evaluation. The problem of evaluating environmental taxes is particularly complex insofar as they are generally applied simultaneously with other instruments (such as regulations), which makes it difficult to isolate the impact of a tax. However, a growing body of data is available, albeit still dispersed.

3.1. Own-price elasticities of demand[1]

3.1.1. Energy products

Most environmentally related taxes apply to the energy and transport sectors. The magnitude of the behavioural responses to environmentally related taxes can be measured in terms of the relevant price elasticities. If, after the introduction of an environmentally related tax, the price of the taxed good increases by 10% and, as a result of the higher price, its consumption falls by 2%, the own-price elasticity in this particular case is –0.2.[2]

Available estimates show that, in most cases, demand for *energy in total* is rather inelastic in the short term; OECD (2000a) show estimates for short run elasticities range between –0.13 to –0.26. However, long run elasticities are considerably higher (–0.37 to –0.46). Nevertheless, an own-price elasticity significantly different from zero indicates that price increases can substantially reduce the demand for energy. Therefore, environmentally related taxes can have a significant impact on reducing energy demand, especially in the long run.[3]

Bjørner and Jensen (2002) used a large panel database with energy data for individual firms to estimate the average energy price elasticity in the industry sector. They found a value of –0.44, which was higher in absolute value than previous Danish estimates based on aggregate time series.[4]

Studies on the price elasticities specifically for *petrol* show comparable, albeit less homogeneous results – as can be seen in Table 3.1.[5] While most estimates show relatively low elasticities in the short run (–0.15 to –0.28), some estimates indicate significantly higher values (–0.51 to –1.07). Long-term elasticities tend to be clearly higher (–0.23 to –1.05). There are differences between countries and variances are to some extent explained by the use of different estimation methods. This leaves policy-makers with certainty about the fact that taxes will have a significant behavioural effect, but uncertainty about the exact magnitude of this effect. This underlines that environmentally related taxes should be implemented in a long-term perspective, avoiding set-back due to political pressures (*e.g.* when world oil prices increase), and with advanced planning and warning of the introduction and/or gradual increase of the tax rates.[6]

Table 3.1. **Selected estimates of own-price elasticities of petrol**

		Short run	Long run	Ambiguous
Pooled time series/ cross section	Micro	−0.30 to −0.39 (USA)	−0.77 to −0.83 (USA)	
	Macro	−0.15 to −0.38 (OECD[1])	−1.05 to −1.4 (OECD[1])	
		−0.15 (Europe	−1.24 (Europe)	
			−0.55 to −0.9 (OECD 18[2])	
		−0.6 (Mexico)	−1.13 to −1.25 (Mexico)	
Cross section	Micro	−0.51 (USA)		
		0 to −0.67 (USA)		
	Macro	Mean −1.07		
		(−0.77 to −1.34) (OECD[1])		
Time series	Macro	−0.12 to −0.17 (USA)	−0.23 to −0.35 (USA)	
Meta-analyses and surveys		Average −0.26 (0 to −1.36) (international)	Average −0.58 (0 to −2.72) (international)	Average −0.53 (−0.02 to −1.59) (USA)
		Mean −0.27 (time series)	Mean −0.71 (time series)	Mean −0.53 (time series)
		Mean −0.28 (cross section)	Mean −0.84 (cross section)	Mean −0.18 (cross section)
				−0.53 (panel data)
				−0.1 to −0.3 (22 estimates)

1. OECD except Luxembourg, Iceland, and New Zealand.
2. OECD 18 covers Canada, the USA, Japan, Austria, Belgium, Denmark, France, Germany, Greece, Ireland, Italy, the Netherlands, Norway, Spain, Sweden, Switzerland, Turkey, and the United Kingdom.
Source: Barde and Braathen (2005), based on OECD (2000a).

Table 3.2. **Selected estimates of own-price elasticities of residential electricity**

		Short run	Long run	Ambiguous
Pooled time series/cross section	Micro	−0.433 (Norway) −0.2 (USA)	−0.442 (Norway)	
	Macro	−0.158 to −0.184 (USA)	−0.263 to −0.329 (USA)	
Cross section	Micro	−0.4 to −1.1 (Norway)	−0.3 to −1.1 (Norway)	
	Macro			−1.42 (53 countries)
Time series	Macro	−0.25 (USA) −0.62 (USA)	−0.5 (USA) −0.6 (USA)	
Meta-analyses and surveys		−0.05 to −0.9	−0.2 to −4.6	−0.05 to −0.12 (4 studies)

Source: OECD (2001a), based on OECD (2000a).

Tables 3.2 and 3.3 present a number of estimates of the own-price elasticity of residential electricity use.[7] While the estimates vary quite a lot, it is clear that increased taxes on electricity would lead to significant reductions in households' demand for electricity.

Holmøy (2005) presents a detailed decomposition of changes in demand for electricity in both industry and households in Norway. He found an own-price elasticity of *aggregate electricity demand* equal to −0.31. Within industry, factor substitution contributed most to this response.

3.1.2. Transport

OECD (2000a) summarized the then available estimates of own-price estimates in the transport sector in Table 3.4.

Table 3.3. **More estimates of own-price elasticities of residential electricity**

Study	Country	Short run	Long run	Details
Aasness and Holtsmark, 1993	Norway		−0.2	Household data
Halvorsen and Larsen, 1998	Norway	−0.33	−0.42	Household data, dynamic model
Parti and Parti, 1980	USA	−0.58		Household data
Morss and Small, 1989	USA	−0.23	−0.38	
Baker, Blundell and Micklewright, 1989	UK		−0.76	Paper includes results for subgroups of households
Dennerlein, 1987	Germany		−0.38	Household data, discrete–continuous choice
Dubin and McFadden, 1984	USA		−0.26	Discrete–continuous choice
Bernard, Bolduc and Bélanger, 1996	Canada	−0.67		Discrete–continuous choice
Branch, 1993	USA	−0.2		Expenditure Survey data
Garbacz, 1983	USA	−0.193		Partial elasticities

Source: Nesbakken (1998).

Goodwin, Dargay and Hanly (2004) include additional information. Table 3.5 summarises their findings regarding studies of the impacts of fuel price changes on various transport demand components when *dynamic estimation methods* – that allow impacts to build up over a defined time period – are used. The long-term elasticities shown here reflect decisions to change place of residence/work in response to higher travel costs. Even longer-term elasticities could be higher again if they also would reflect changes in urban development patterns in response to higher transport costs. Table 3.6 presents the survey by Goodwin, Dargay and Hanly (2004) of estimation studies of the impacts of fuel price changes when *static estimation techniques* are used – where the estimates are meant to capture impacts at an undefined future time when all responses to the price changes have been completed. Table 3.7 shows results of studies that have estimated similar types of impacts of a change incar purchase costs.

Table 3.4. **Own-price elasticities of modes of transport**

	Short run	Long run	Ambiguous
Elasticity for automobiles	−0.09 to −0.24	−0.22 to −0.31	−0.13 to −0.52
Elasticity for urban transit			
Time-series			−0.01 to −1.32[1]
Cross-section			−0.05 to −0.34
Pooled data			−0.06 to −0.44

	Time-series	Cross-section
Elasticity for air travel		
Leisure	−0.4 to −1.98	−1.52
Business	−0.65	−1.15
Mixed or Unknown	−0.36 to −1.81	−0.76 to −4.51
Elasticity for inter-city rail		
Leisure	−0.67 to −1.00	−0.7
Mixed or unknown	−0.37 to −1.54	−1.4

1. Most values fall between −0.1 and −0.6.
Source: OECD (2000a).

Tables 3.5-3.7 include both own-price and cross-price elasticities. From the cross-price elasticities one can see that a higher fuel price will tend to reduce the size of the car stock

Table 3.5. **Elasticities of various measures of transport demand with respect to fuel price per litre**

Produced by dynamic estimations, using time series data

Dependent variable	Short-term	Long-term
Fuel consumption (total)		
Mean elasticity	−0.25	−0.64
Standard deviation	0.15	0.44
Range	−0.01, −0.57	0, −1.81
Number of estimates	46	51
Fuel consumption (per vehicle)		
Mean elasticity	−0.08	−1.1
Standard deviation	n.a.	n.a.
Range	−0.08, −0.08	−1.1, −1.1
Number of estimates	1	1
Vehicle-km (total)		
Mean elasticity	−0.10	−0.29
Standard deviation	0.06	0.29
Range		−0.63, −0.10
Number of estimates	3	3
Vehicle-km (per vehicle)		
Mean elasticity	−0.10	−0.30
Standard deviation	0.06	0.23
Range	−0.14, −0.06	−0.55, −0.11
Number of estimates	2	3
Vehicle stock		
Mean elasticity	−0.08	−0.25
Standard deviation	0.06	0.17
Range	−0.21, −0.02	−0.63, −0.10
Number of estimates	8	8

Source: Goodwin, Dargay and Hanly (2004). n.a. = Not available.

to some extent, in particular over the long term. Likewise, a higher cost of purchasing a car will reduce both the fuel consumption and the total number of km driven quite significantly – especially in the long term.[8]

3.1.3. Waste

Fullerton (2005) includes the estimates of impacts of variable waste collection charges reproduced in Table 3.8. With a notable exception for the study by Podolsky and Spiegel, the change in (unsorted) garbage stemming from a price increase is found to be quite modest.[9]

Dijkgraaf (2004) analysed various waste collection payment systems in the Netherlands and compared the price elasticities of different systems of variable collection charges. The results are summarised in Table 3.9 – for an estimation model that includes a variable to correct for the possibility that the municipalities that have introduced variable collection charges are characterised by a particularly high degree of environmental activism. Even after having made this modification, the estimated elasticities for some of the payment schemes are rather high compared to those presented in Fullerton (2005).

It is worth noting that the price elasticities found for a collection charge system with payments per bag for both unsorted and compostable waste are almost as high as the elasticities of a charging system based on weighted waste amounts. This can be of importance, as the administrative costs of a payment-per-bag system are likely to be significantly lower than the costs of running a weight-based system.

Table 3.6. **Elasticities of various measures of transport demand with respect to fuel price per litre**

Produced by static estimations

Dependent variable	Total	Cross-section data	Cross-section /time series data	Time series data
Fuel consumption (total)				
Mean elasticity	−0.43	−0.55	−0.28	−0.48
Standard deviation	0.23	0.32	0.10	0.16
Range	−0.11, −1.12	−0.23, −1.12	−0.45, −0.11	−0.77, −0.28
Number of estimates	24	7	9	8
Fuel consumption (per vehicle)				
Mean elasticity	−0.30	No observations	−0.30	No observations
Standard deviation	0.22		0.22	
Range	−0.89, −0.04		−0.89, −0.04	
Number of estimates	22		22	
Vehicle-km (total)				
Mean elasticity	−0.31	−0.38	−0.27	−0.32
Standard deviation	0.14	0.23	0.12	–
Range	−0.54, −0.13	−0.54, −0.21	−0.41, −0.13	−0.32, −0.32
Number of estimates	7	2	4	1
Vehicle-km (per vehicle)				
Mean elasticity	−0.51	No observations	−0.33	−0.69
Standard deviation	0.25		–	–
Range	−0.69, −0.33		−0.33, −0.33	−0.69, −0.69
Number of estimates	2		1	1
Vehicle stock				
Mean elasticity	−0.06	0.03	−0.11	No observations
Standard deviation	0.08	–	0.03	
Range	−0.13, 0.03	0.03, 0.03	−0.13, −0.09	
Number of estimates	3	1	2	

Source: Goodwin, Dargay and Hanly (2004).

Table 3.7. **Elasticities of various measures of demand with respect to car purchase costs**

Whole database

Dependent variable	Short-term	Long-term	Static
Fuel consumption (total)			
Mean elasticity	−0.12	−0.51	−0.45
Standard deviation	0.08	0.24	0.25
Range	−0.26, 0.00	−0.88, 0.00	−0.66, −0.15
Number of estimates	11	10	4
Vehicle-km (total)			
Mean elasticity	−0.19	−0.42	−0.35
Standard deviation	0.12	0.21	0.42
Range	−0.33, 0.11	−0.62, −0.20	−0.65, −0.05
Number of estimates	3	3	2
Vehicle stock			
Mean elasticity	−0.24	−0.49	−0.38
Standard deviation	0.15	0.19	0.29
Range	−0.44, −0.03	−0.78, −0.13	−0.59, −0.05
Number of estimates	11	11	3

Source: Source: Goodwin, Dargay and Hanly (2004).

Table 3.8. **Empirical estimates of the effect of unit pricing for waste collection**

Study	Data	Model	Change in garbage	Change in recycling
Wertz (1976)	Compares subscription program vs. flat fee	Comparison of means	$\varepsilon = -0.15$	–
Jenkins (1993)	Panel of 14 cities (10 with user fee), 1980-88	GLS	$\varepsilon = -0.12$	–
Hong, Adams and Love (1993)	1990 survey of 4 306 homes in Portland, OR	Ordered probit, 2SLS	No significant impact	Positive relationship
Reschovsky and Stone (1994)	1992 mail survey of 1 422 households in and around Ithaca, NY	Probit	–	No significant impact
Miranda et al. (1994)	Panel of 21 cities over 18 months starting in 1990	Comparison of means	17%-74% less garbage	Increase of 128%
Callan and Thomas (1997)	1994 cross-section of 324 towns in MA, 55 with unit-pricing programs	OLS	–	6.6-12.1% point increase
Fullerton and Kinnaman (1996)	Two-period panel of 75 households in 1992, in Charlottesville, VA	OLS	$\varepsilon = -0.076$ (weight), or -0.23 (cans)	Cross-price elasticity is 0.073
Van Houtven and Morris (1999)	Monthly panel for 400 households in Marietta, GA, in 1994	Random effects model	-36% for bags, -14% for cans	No significant impact
Podolsky and Spiegel (1998)	1992 cross-section of 159 municipalities in NJ, 12 with unit-pricing	OLS	$\varepsilon = -0.39$	–
Kinnaman and Fullerton (2000)	1991 cross-section of 959 towns across the US, 114 with unit-pricing	OLS 2SLS	$\varepsilon = -0.19$ $\varepsilon = -0.28$	$\varepsilon = 0.23$ $\varepsilon = 0.22$

ε = price elasticity of demand; OLS = ordinary least squares, GLS = generalised least squares, and 2SLS = two-stage least squares.
Source: Fullerton (2005).

Table 3.9. **Price elasticities of different variable waste collection systems in the Netherlands**

System	Average price	Total waste	Unsorted waste	Compostable waste	Recycled waste
Payments according to weight	4.39	−0.40	−0.53	−0.81	0.12
Payments per bag for unsorted and compostable waste	2.02	−0.36	−0.51	−0.85	0.20
Payments per bag for unsorted waste	2.15	−0.07	−0.58	0.40	0.09
Payments according to collection frequency	3.91	−0.16	−0.16	−0.31	0.04
Payments according to waste volume	1.94	−0.00	0.01	0.09	−0.03

Source: Dijkgraaf (2004).

3.1.4. Pesticides

As another example not related to energy use, available estimates indicate an own price elasticity of pesticide use in the range of –0.2 to –1.1, see Table 3.10. The estimates are in most cases well below –1.0 in absolute value – meaning that the demand for pesticides in economic terms is described as "inelastic". However, the available estimates are clearly different from 0, meaning that a tax on pesticides should clearly contribute to lower pesticide use.

Table 3.10. **Estimates of own-price elasticities of pesticides.**

Study	Country	Elasticity	Remarks
Oskam (1997)	EU	–0.2 to –0.5	General overview of other studies
DHV and LUW (1991)	Netherlands	–0.2 arable farming –0.3 horticulture	Short term
Oskam (1992)	Netherlands	–0.1 mixed farms –0.5 specialised farms	Medium term
Oude Lansink and Peerlings (1995)	Netherlands	–0.5 –0.7 (with CAP reform)	Based on data 1970-1992
Russell (1995)	UK	–1.1	Based on 26 cereals producers, 1989-93
Falconer (1997)	UK	–0.3	Using a linear programming model
Ecotec (1997)	UK	–0.5 to –0.7	Only for herbicides used for cereal grass weed
Dubgaard (1987)	Denmark	–0.3	Using a threshold model
Dubgaard (1991)	Denmark	–0.7 herbicides –0.8 fungicides and insecticides	Period 1971-1985
Schulze (1983)	Germany	–0.5 fungicides	Using a linear programming model
Johnsson (1991)	Sweden	–0.3 insecticides, –0.4 fungicides	Based on filed experiments
Gren (1994)	Sweden	–0.4 fungicides –0.5 insecticides, –0.9 herbicides	Econometric model
SEPA (1997)	Sweden	–0.2 to –0.4	General overview
Rude	Norway	–0.2 to –0.3	
Carpentier	France	–0.3	Arable farms
Papanagiotou (1995)	Greece	–0.28	

Source: Hoevenagel *et al.* (1999) and Muños Piña (2004).

3.2. Cross-price elasticities

As mentioned above, the cross-price elasticities (*e.g.*, between different fuels) are also of importance for the environmental effectiveness of a tax. This is to some extent shown by the rapid switch from high-sulphur to low-sulphur petrol and diesel in United Kingdom illustrated in Figure 2.8 above – although that resulted more from changes in supply than from changes in demand.

The more gradual, but still very significant, change in the composition of motor fuels used – from petrol to diesel – especially in Europe, is probably a better case in point, reflecting *i.a.* the lower taxes on diesel than on petrol in most of these countries.

3.3. Examples of available estimates of changes in demand

3.3.1. Changes in the demand for petrol and diesel at a European level

As indicated in Section 2.2, the small decline in revenues from environmentally related taxes measured in per cent of GDP since 1999 can to a certain extent be explained by a reduction in demand for petrol in OECD Europe. Figure 3.1 illustrates some relevant developments.

Between 1994 and 2000, the weighted average nominal tax rate on petrol in Europe increased almost 50%, from EUR 0.5 to 0.72 per litre.[10] Between 1994 and 1999, the import price of petrol was more or less constant, but between 1999 and 2000 a strong increase took place – in response to the similar increase in world crude oil prices. The tax increase and the higher import price of petrol lead to a significant increase in user-prices of petrol. Hence, the sales of (the highly taxed) petrol in OECD Europe peaked in 1999, and a *significant decrease* has since been observed. By 2004 a 10% reduction compared to 1999

Figure 3.1. **Sales of, and taxes on, petrol and diesel in OECD Europe**
1994-2004

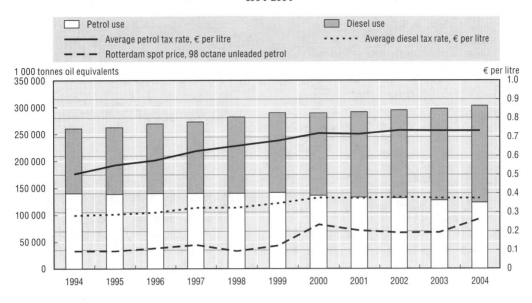

Source: IEA databases and OECD estimates. The average tax rates have been calculated by weighing IEA data on petrol and diesel taxes (including VAT) paid by households with the share of each country in the total use of petrol and diesel in OECD Europe. The Rotterdam spot price for diesel is very close to the spot price for petrol shown in the figure. The 2004 numbers for petrol and diesel use are preliminary figures.

could be seen. The user-price of diesel did also increase in this period (even if the tax increases were less important) – but from a considerably lower starting point than for petrol, as the average tax rate is only about the half of that for petrol. Car users in Europe thus had a strong, and increasing, incentive to switch from petrol-driven cars to diesel-driven cars. Whereas petrol represented 54% of the combined petrol and diesel use *in the whole transport sector* in 1994, this share had fallen to 43% in 2003.[11]

As an *illustration*, the revenue "foregone" by the 10% reduction in petrol sales in OECD Europe between 1999 and 2003 would represent some EUR 10.8 billion at an assumed net-of-VAT tax rate of EUR 0.6 per litre – or approximately 0.04% of GDP of *all* OECD member countries – if the petrol not bought was *not* replaced by diesel. If all the petrol *was* replaced by diesel, which is taxed at approximately half the rate per litre, the revenue "lost" could represent some 0.02% of the GDP of *all* OECD member countries. In both cases, this "explains" a significant part of the reduction in average revenues from environmentally related taxes in per cent of GDP between 1999 and 2003 that could be seen in Figure 2.1.

In other parts of OECD, an outright reduction in petrol (or diesel) sales did not take place between 1999 and 2003. This could in part be because exchange rate developments contributed to a much slower increase in the user prices of the fuels in the years leading up to 2000 than in European countries. However, as exemplified in Figure 3.2, fuel use has *grown slower than real GDP* in all OECD regions. This also contributes to reducing the revenues from environmentally taxes measured in per cent of GDP.

Figure 3.2 show similar information as Figure 3.1, for a few selected countries, namely Germany, Norway, Turkey and United States.[12] However, to enhance comparability between countries, the sales of petrol and diesel measured in tonnes oil equivalents has been divided by real GDP measured in USD, converted using purchasing power parities

Figure 3.2. **Sales of, and taxes on, petrol and diesel in selected countries**

Sales per unit of GDP, converted to USD using 2000 purchasing parities, 1994-2004

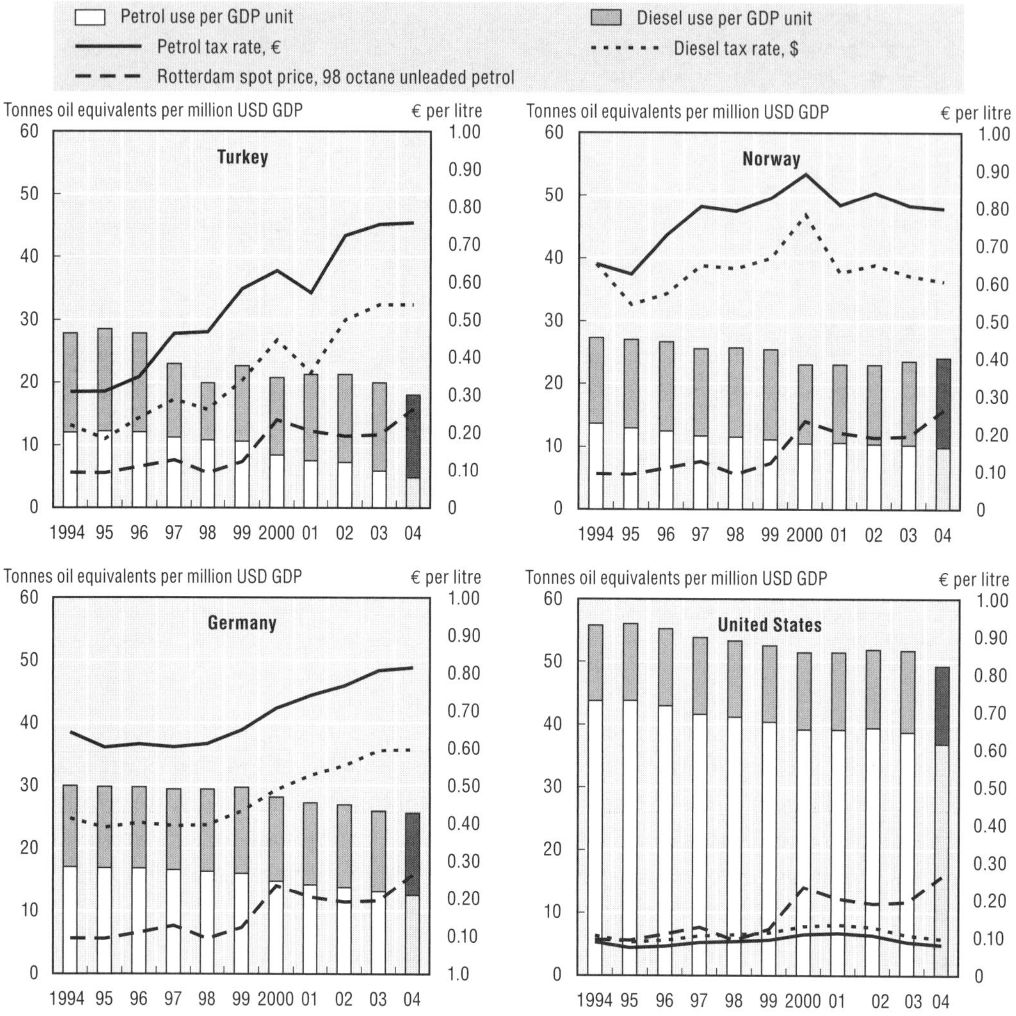

Source: IEA databases and OECD estimates. The variations over time in the tax rates in United States only reflect changes in the exchange rate between dollars and euros. Both federal and state fuel taxes are included. The 2004 numbers for petrol and diesel use are preliminary figures.

from 2000. Several points can be made. The significant tax increases in Germany and Turkey have – in combination with the crude oil price increases – contributed to a marked *decrease* in the sale of petrol, but also of motor fuels in total, in both these countries.[13] In Norway, where tax rates were decreased in the wake of the crude price increases in 2000, the sale of petrol and diesel has changed less in recent years. In United States, the comparatively low fuel taxes has contributed to a much higher use of petrol and diesel per unit of GDP than in the other countries. One can, however, notice the much less important role played by diesel in United States than in the 3 other countries depicted in the graph – in part because the tax rates on diesel are slightly higher than the rates for petrol.

The developments outlined here emphasise the important role price-based instruments can play – even for products that are relatively "inelastic" in demand, like petrol and diesel. They also highlight the role cross-price elasticities can play, illustrated by the switch from petrol to diesel, in particular in OECD Europe – where the difference in taxes on petrol and diesel is the largest.

3.3.2. Examples of studies made in some OECD member countries

As experience in environmental taxes grows, an increasing body, albeit still limited, of estimates of changes in demand becomes available. A few examples are presented below.

There are several estimates relating to the impacts of the CO_2 tax in *Denmark*.[14] According to Nordic Council of Ministers (2002), CO_2 emissions in Denmark decreased 6% during the period 1988-1997 while the economy grew by 20%. They also decreased 5% just between 1996 and 1997, when the tax rate was raised.

According to Schou (2005), the introduction of the pesticides tax in *Denmark* in 1996 contributed to a reduction in pesticides use by some 10-13% from 1995 to 1996 (although other factors can also have influenced the development). A doubling of the tax rates in 1998 contributed to a reduction in the treatment frequency from 2.45 to 2.10 from 1999 to 2002.[15]

According to the Ministry of Taxation (2002), the *differentiation* of the tax rates on diesel according to the sulphur content of the fuel helped reduce SO_2 emissions by 6,550 tonnes in 2000. The economic value of this emission reduction was estimated to DKK 373 million. In addition to this, the tax on diesel fuels as such was estimated to have reduced the SO_2 emissions by approximately 775 tonnes, with an economic value of DKK 44 million.

To reveal the driving forces behind the changes in *Norwegian* emissions of the three most important climate gases, CO_2, methane and N_2O in the period 1990-1999, Bruvoll and Larsen (2004) decomposed the actually observed emissions changes, and used an applied general equilibrium simulation to look into the specific effect of carbon taxes. Although total emissions did increase, they found a significant reduction in emissions per unit of GDP over the period due to reduced energy intensity, changes in the energy mix and reduced process emissions. Despite considerable taxes and price increases for some fuel-types, the effect of the carbon tax was, however, modest. While the partial effect from lower energy intensity and energy mix changes was a reduction in CO_2 emissions of 14%, the carbon taxes contributed to only a 2% emission reduction. This relatively small effect can – according to the authors – be explained by extensive tax exemptions and relatively inelastic demand in the sectors in which the tax is actually implemented.

Sweden introduced in 1992 a charge on measured NO_x from combustion plants over a certain size. The revenues raised are repaid to the firms affected according to the amount of energy they produce.[16] Seen as a group, the plants are – in other words – not much affected by the charge payments. As shown in Figure 3.3, there are, however, large differences between the production units – in all of the sectors affected by the charge. The most energy-efficient units in each sector receive a net payment from the charge-and-refund mechanism, while the least energy-efficient units are net losers. One can notice that the pulp and paper industry and the wood industry both are net losers from the scheme. Sterner and Höglund Isaksson (2006) state that it is perhaps characteristic that the only suggestion for reform of the scheme has come from the pulp and paper sector – who wanted the refunded charge to still be used, but with a separate refunding for the plants within each industry.

This charge on NO_x has been quite effective in reducing emissions from the combustion plants that it covers – in part because the rate is quite high, at EUR 4.4 per kg NO_x emitted.[17] When the charge was introduced, only plants producing more than 50 GWh per year were included, but this limit was reduced to 40 GWh in 1995, and further down to 25 GWh in 1996. Figure 3.4 illustrates both total emissions from the (increasing number of) plants covered, and the emissions per MWh energy produced. The emissions per unit energy produced are now less than a half of what they were in 1990, before the charge was introduced.[18]

Figure 3.3. **Net payers and receivers in the relation to the refunded NO$_x$ charge in Sweden**

Production units in different industrial sectors, 2004

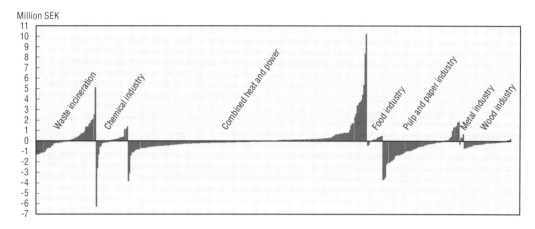

Source: Swedish Environmental Protection Agency.

Figure 3.4. **Total and specific NO$_x$ emissions from energy-producing**

The plants covered by the NO$_x$ charge in Sweden, 1990-2004

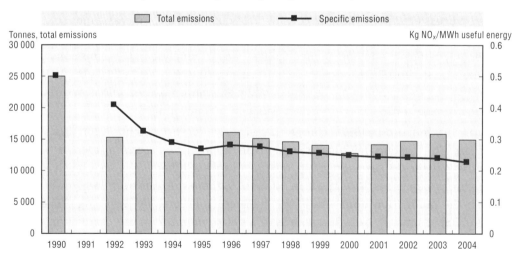

Source: Swedish Environment Protection Agency

The Swedish sulphur tax (introduced in 1991) led to a fall in the sulphur content of oil-based fuels of more than 50% beyond the legal standards. The sulphur content of light oils has now fallen below 0.076% (i.e. less than half the legal limit of 0.2%). The tax is estimated to have reduced emissions of sulphur dioxide by 80% compared to 1980 (Nordic Council of Ministers 1999).

Several studies made for the Federal Environment Agency in *Germany* on the overall social impacts of the ecological tax reform identified clear signs of the desired ecological effects. Energy consumption has been decreasing and CO_2 emissions could be reduced by 2-3% by 2005 – compared to what would have taken place without the reform in place.[19]

There have also been important environmental benefits stemming from tax incentives given in Germany to the purchase of vehicles fulfilling the Euro 3 and Euro 4 emission norms – before the fulfilment of these norms became compulsory at the EU level. As can be seen from Figure 3.5, the market share of vehicles fulfilling these norms increased rapidly once the tax incentives were introduced.[20]

Figure 3.5. **Share of vehicles fulfilling the Euro 3 and Euro 4 norms**

In total sales of new passenger vehicles in Germany, 1997-2004

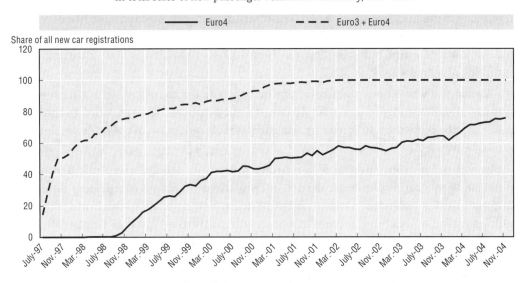

Notes: Between 1.7.1997 and 31.12.1999 those who had bought a passenger vehicle that fulfilled the so-called Euro 3 emission norm could get a cumulative reduction in the annual tax on limited upwards to approximately 125 € for petrol-driven vehicles and 250 € for diesel-driven vehicles.

Between 1.7.1997 and 31.12.2005, a similar cumulative tax reduction was available to those that bought vehicles that fulfilled the stricter Euro 4 norm, limited upwards to about 300 € for petrol-driven vehicles and 600 € for diesel-driven ones.

The tax in which the reductions were given depended on the cylinder volume of each vehicle, and the tax rates were higher for diesel-driven vehicles than for petrol-driven ones. A car with a small cylinder volume would thus remain exempt from taxation over a longer time period than a car a large engine – but no cars were exempted beyond 31.12.05, even if the owner had not yet benefited from the maximum amount of tax reduction under the scheme. As from 01.01.06, all new vehicles must fulfil the Euro 4 norm.

Source: Federal ministry for the environment, nature conservation and nuclear safety, Germany.

Cambridge Econometrics (2005) presents an in-depth analysis of the impacts of the Climate Change Levy in the *United Kingdom*, comparing actual emission developments to a counterfactual reference case with no levy in place and estimating developments up to 2010 under various assumptions. The study *inter alia* found that total CO_2 emissions were reduced by 3.1 mtC (million tonnes carbon) – or 2.0% – in 2002 and by 3.6 mtC in 2003 compared to the reference case. The reduction is estimated to grow to 3.7 mtC – or 2.3% – in 2010. Most of the reduction (1.8 mtC in 2010) was found to take place among "other final users", *i.e.* in commerce and the public sector, but "other industry" – *i.e.* industry other than basic metals, mineral products and chemicals – was also found to reduce emissions around 0.8 mtC in 2010. Emissions from power generation were also found to decrease, due to lower demand for electricity. When interpreting these results it should be kept in mind that households are exempted from the Climate Change Levy, and that energy-intensive industries can benefit from an 80% tax rate reduction if they fulfil negotiated energy efficiency targets.[21]

The aggregates tax in United Kingdom, introduced in April 2002 on the extraction of *e.g.* rock, sand and gravel, has also been found to be effective. According to HM Treasury (2005), sales of primary aggregate in Great Britain fell by 8% between 2001 and 2003, against a backdrop of buoyant construction activity and GDP growth, ascan be seen in Figure 3.6.

HM Treasury (2005) also contains an evaluation of the Landfill Tax in the United Kingdom. The quantity of inactive or inert waste disposed to landfill fell by 60% between 1997-98 and 2003-04. Allowing for the fact that some of this material may have been reclassified as exempt, there is still an overall reduction in inert or exempt material of 16 million tonnes (35%) over the period. The total amount of waste landfilled has decreased less, as the reduction in the landfilling of active (biodegradable) waste has been smaller than for inactive waste, but – as shown in Figure 3.7 – a reduction is now occurring in the landfilling of active waste.[22]

Also in the waste area, the tax on plastic bags applied in *Ireland* since 2002 seems to have had significant environmental impacts. It has contributed to a reduction in the use of plastic bags by more than 90%, leading to a considerable reduction of the litter problem, see Box 8.1 below.

In most countries, the tax differentiation between leaded and unleaded petrol, combined with a series of measures such as regulations making it compulsory for service stations to offer unleaded petrol and introducing new emission standards for motor vehicles (based on such requirements as catalytic converters) led to a strong fall in the market share of leaded petrol, which is now withdrawn from sale in almost all OECD

Figure 3.6. **Sales of aggregates, real construction output and real GDP in the United Kingdom**

1982-2004

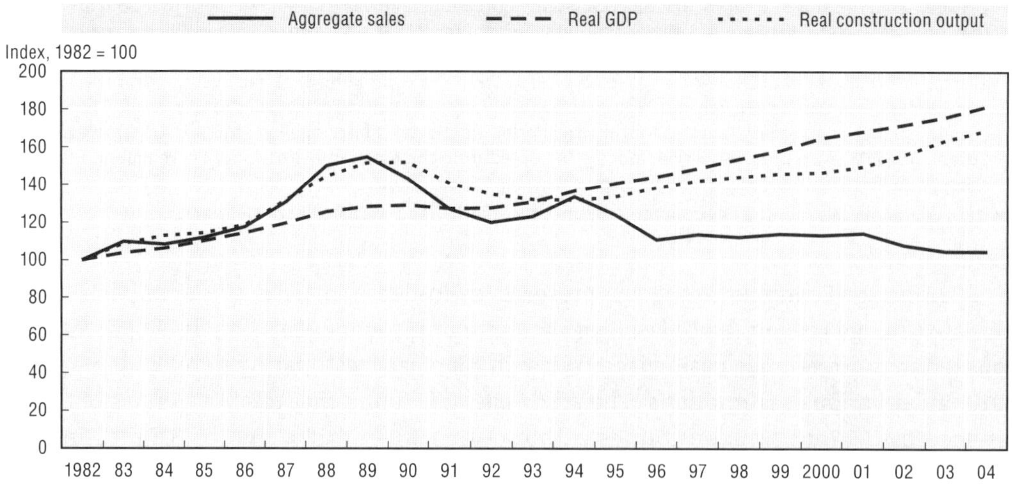

Source: British Geological Survey, UK Department of Trade and Industry, UK Office for National Statistics.

Figure 3.7. **Landfilling of active waste and the standard tax rate of the Landfill Tax**
The United Kingdom, 1997-98 to 2003-04

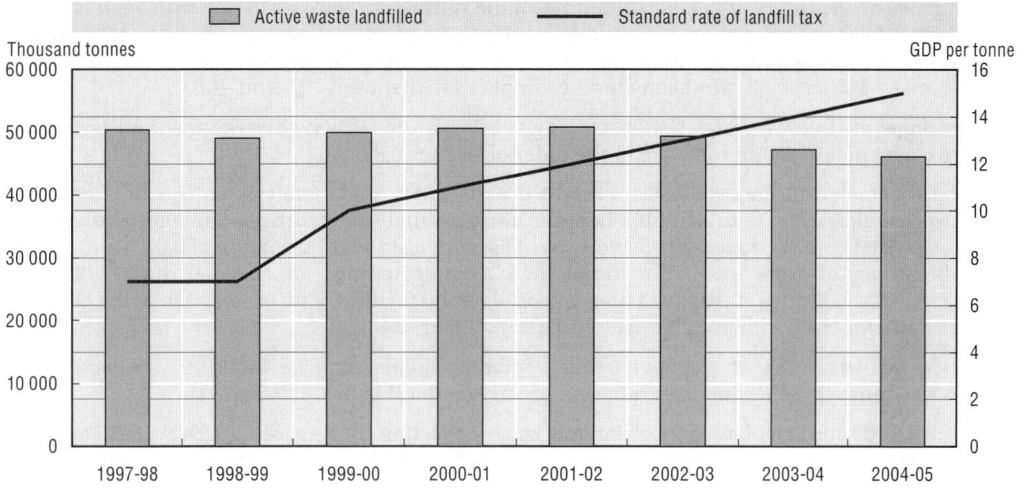

Source: HM Revenue & Customs.

countries. The fiscal incentive greatly speeded up the process, despite slow penetration of new vehicles equipped with catalytic converters.

In *France*, Salanié and Thomas (1999) have assessed the effects of the water pollution tax, levied by regional water agencies. The tax rate varied both across time (between 1985 and 1995) and between the four water agencies, and the authors estimated the price elasticity to be between –0.7 and –0.8.

Not all taxes have been successful, however. The rates of French tax on SO_2 emissions (EUR 38 per tonne) – only a fraction of similar rates in the Nordic countries – are, according to Riedinger (2005), well below the cheapest abatement measures industry could take. The tax has, hence, had minimal impact on firms' behaviour.[23] Similarly, the rates of the Swedish tax on pesticides were, according to Swedish Environmental Protection Agency (1997), too low to produce incentive effects.

Notes

1. A comment on different approaches to estimating price elasticities is given in the Technical Annex at the end of the publication.

2. For more information see OECD (2000a, 2001a) and Barde and Braathen (2005).

3. Liu (2004) presents additional estimates for OECD countries.

4. They also found that the price elasticities depended on the level of the energy prices firms were facing at the outset. For firms at the 10% decile when ranked in increasing order according to energy prices they were facing, the estimated price elasticity of energy was about –0.4. For firms at the median, the price elasticity was found to be about –0.6, while for firms at the 90% decile, the estimated price elasticity of energy was about –0.7.
   To the extent that the sectoral competitiveness arguments often used in favour of special provisions for energy-intensive firms are valid, infra-marginal price increases due to major tax increases could, however, also trigger plant closures that (obviously) would eliminate energy use at a given plant. It is doubtful that the price elasticity estimates presented above incorporate such impacts, as energy-intensive firms in Denmark (as elsewhere) have enjoyed special tax privileges

all through the estimation period. Expected demand reductions in response to significant tax increases could, hence, be higher than what the presented estimate indicate.

5. Results of a survey of price elasticity estimates undertaken among ministries of finance and ministries of environment in OECD countries in the summer of 2005 are in line with what is shown in Table 2.4: Estimates of short-term price elasticities for petrol (reflecting demand changes taking place within one year) are in the range –0.15 to –0.3, medium term estimates (up to 10 years) range between –0.35 and –0.52, while long term estimates vary between –0.8 and –1.0.

6. Hautzinger *et al.* (2004) states that, according to a literature survey, short- to medium-term own-price elasticities of motor fuels are in the order of –0.2 to –0.3.

7. Bye, Langmoen and Aasness (2004) present a preliminary meta-analysis of 27 estimates of the own-price elasticity of households demand for electricity in Denmark, Finland, Norway and Sweden. They found an average elasticity of –0.53, with a standard deviation of 0.08.
In the survey of price elasticity estimates OECD undertook in the summer of 2005, Belgium indicated an own-price elasticity of households' demand for electricity of –0.19, while the elasticity of demand for electricity in other uses was estimated to –0.40.

8. Many other estimates of transport-related price elasticities can be found on the web-site of Victoria Transport Policy Institute, at *www.vtpi.org/tdm/tdm11.htm#_Toc68662033*.

9. For further discussion of impacts of variable waste collection charges, see OECD (2006a). Bartelings *et al.* (2005) provide a brief overview of studies of impacts of landfill taxes and various waste collection schemes.

10. The average tax rates have been calculated by weighing (mostly) IEA data on petrol and diesel taxes paid by households (these numbers *include* VAT) with the share of each country in the total use of petrol and diesel in OECD Europe.

11. These numbers for the transport sector as a whole include *e.g.* (diesel) fuels used by heavy goods vehicles. The shift in fuel composition from petrol to diesel among passenger cars and similar has thus been stronger than these percentages could indicate.

12. VAT is included in the tax rates shown here – as opposed to what is the case in Figure 2.6.

13. To the extent that any "tank tourism" has taken place – i.e. that car users (to an increased extent) have bought fuels in neighbouring countries with lower tax rates – the net environmental improvements caused by the tax increases can have been smaller than the impression given by Figure 3.2. In this case, developments in fuel *sales* would be an imprecise proxy for changes in fuel *use* in the country in question. Whereas "tank tourism", for geographical reasons, can seem to be of limited relevance for most car users in Turkey, it could be of more importance in the case of Germany.

14. See Nordic Council of Ministers (2001 and 2002) for additional information.

15. Treatment frequency expresses the number of times the total area of arable land can be treated on average with the sold quantities of pesticides, when they are used at the normal dose rates.

16. For more information, see Swedish Environmental Protection Agency (2000), Naturvårdsverket (2003), Höglund Isaksson (2005) and Sterner and Höglund Isaksson (2006).

17. As a comparison, the tax rate on measured NO_x emissions from waste incinerators in Norway is EUR 1.8 per kg – based on estimates of the social costs of such emissions. The revenues from the Norwegian tax are, however, not repaid to the incinerators affected. In France, a non-refunded tax of EUR 0.04573 per kg NO_x emitted is levied as part of the "Taxe générale sur les activités polluantes", the General tax on polluting activities.

18. A side effect of the NO_x charge is higher emissions of, among other things, carbon monoxide and N_2O, which are not regulated by the charge.

19. See UBA (2004), Bach (2005), Knigge and Görlach (2005a and 2005b) and Kohlhaas (2005) for further information.

20. From the perspective that polluters ought to pay for the harm they cause, it could have been more appropriate to increase the tax on vehicles not fulfilling the respective Euro norms than to give subsidies to the vehicles that did fulfil them – also because the subsidies would contribute (marginally) to an increase in the total number of vehicles. While it is not quite clear how the market share of the relatively clean vehicles would have developed in the absence of the subsidies given, Figure 3.5 nevertheless gives a clear illustration of the impact the price incentives can have on the composition of the demand for motor vehicles.

21. For further discussion of the Climate Change Levy, see Sections 6.2 and 10.5.3 below.

22. While the tax rate for inactive waste (*e.g.* construction and demolition waste) has been kept at its original level of GBP 2 per tonne ever since the Landfill Tax was introduced in 1999, the tax rate for active waste (*e.g.* unsorted waste from households) has been steadily increasing – and is set to increase significantly further in the years to come, up to a medium-term level of GBP 35 per tonne. According to HM Treasury (2006), provisional figures show that between 1997 and 2005, the volume of active waste disposed at landfill sites fell by almost 16%, with the biggest fall occurring in the last year. See OECD (Forthcoming) for further discussion of the UK Landfill Tax.

23. See also Conseil des impôts (2005). The marginal abatement costs for SO_2 emissions were estimated to be between EUR 310 and EUR 990 per tonne. The "Taxe générale sur les activités polluantes" (TGAP) (General tax on polluting activities) is also – among others – levied on emissions of NO_x (EUR 45.73 per tonne), N_2O (EUR 57.17 per tonne), Hydrochloric acid (EUR 38.11 per tonne) and VOC (EUR 38.11 per tonne).

ISBN 92-64-02552-9
The Political Enconomy of Environmentally Related Taxes
© OECD 2006

Chapter 4

The Sectoral Competitiveness Issue – Theoretical Studies

4.1. Introduction

As underlined in OECD (2001a), a major obstacle to the implementation of environmentally related taxes in certain cases is the fear of reduced international competitiveness in the most affected economic sectors. This concern is not only for economic but also for environmental reasons. In general, from an economic point of view, when introducing of environmentally related taxes increases domestic production costs of internationally traded goods, domestic production generally would be expected to decline – exports become less attractive and imports more – at least in the short run, implying job-losses and other adjustments in the national economy.

Additionally, competitiveness concerns are also a key issue for environmental reasons. When the introduction of environmentally related taxes may result in significant reduction in profits for some sectors, their implementation would be inefficient in terms of environmental improvement if: i) producers relocate operations to places where similar taxes are not implemented (not-controlled regions); or ii) consumers buy more goods from not-controlled regions. Consequently, it could be expected that non-unilateral implementation actions will imply, at least, better environmental results for example in the reduction of CO_2 emissions, as explained in the following section.

There are three key issues when assessing competitiveness:

- how competitiveness is defined;
- the *ex-ante* and *ex-post* scenarios; *i.e.*, the "baseline" policy against which the impact is being assessed;
- the environmental policy instrument that is used.

We will discuss these three issues before evaluating the threat of competitiveness using two theoretical case studies, simulations in the steel and cement manufacturing sectors.

4.1.1. Definition of competitiveness

A first point to underline is that any policy instrument used to achieve environmental targets should cause changes in consumption and/or production patterns. If a policy fails to create such changes it simply cannot deliver any environmental improvements. The more relevant issues are, hence, who should change their behaviour, by how much and within which timeframe.[1]

It is also important to differentiate between the different levels of the competitiveness concept, particularly between the competitiveness of individual firms and sectors and the whole economy of a country, as underlined in OECD (2003b). A company is competitive if it is able to produce products that are either cheaper or better than those of other firms. Ultimately, business competitiveness is a matter of relative performance. The impact of environmental regulation may be complex, and may well vary between firms.[2]

Applying the concept of competitiveness to industrial sectors or to whole economies is more controversial. At a national level, any negative impact imposed upon one firm or sector will tend to be attenuated by positive impacts on others. In other words, at the economy-wide level, correcting for market failures provides an improvement in the overall economic outcome, and what represents increased costs for one firm, sector or industry may lead to reduced costs for others. A prime example is the introduction of higher energy taxes when the revenue is recycled through lowering social security contributions. In this case the competitiveness of labour intensive production will improve.

While there are many good reasons why policy makers ought to focus more upon impacts of (environmental) policies at a national than at a sectoral level, in practise they tend to be more concerned with any potential "losers" from a policy change than with the impacts on the economy as a whole.[3] Hence, the focus of the present discussion is on competitiveness impacts at a sectoral level – *not* at a national level. Considerable attention is given to policies addressing climate change, in part because such policies are likely to trigger more significant behavioural changes throughout the economy than most other environmental policies and in part because of the problem of "leakage". The latter arises when the unilateral imposition of environmentally related taxes in one country results in the relocation of production to other countries. In the case of taxes levied on "local" pollutants, the loss of competitiveness as evidenced by the relocation may be judged to be worthwhile – because of the resulting local environmental improvement. However, when the pollutants concerned contribute to global problems, the loss of competitiveness in the country imposing the tax results in little or no local environmental improvement, as the country continues to suffer from the pollution even though the activities that produce it have move abroad.

There are mainly three determinant factors driving sectoral competitiveness when imposing an environmentally related tax. Thus, the effects on competitiveness will be stronger:

1. The lower the *ability to pass on costs increases in prices*. This will depend on the price-responsiveness of demand, the market structure (number of players, state involvement – regulation or state ownership) and the geography of the sector market; international competition being the most important factor in reducing this ability.[4]

2. The lower the *feasibility of the substitution possibilities*; as limited scope for identifying and financing cleaner production technologies and processes implies an inability to substitute away from environmental taxes.

3. The higher the *energy intensity of the sector*, since the bulk of environmentally related taxes is levied on energy use and transportation.

4.1.2. Competitiveness assessment and "baseline" policy

When assessing the impact of environmentally related taxes on competitiveness it is important to clearly specify the alternative "baseline" policy against which the impact is being assessed. Two dimensions are particularly relevant:

- The *impact on the government budget*: whether the comparison is done on a revenue-neutral basis, or tax revenues are assumed to be higher with the environmental tax than in the baseline case. Generally, comparisons need to be made on a revenue-neutral basis, or else the effects will be substantially complicated (and perhaps dominated) by the macroeconomic impact of the change in the government deficit.

- The *impact on the environment*: whether the comparison is between two equivalent ways of achieving a given standard of environmental protection, or the level of environmental protection vary between the two scenarios.

For example, an environmentally related tax on domestic industrial emissions can be introduced in a revenue neutral basis by using its tax revenues to replace existing distortionary taxes such as social security contributions. The implementation of such a policy will raise three main effects on competitiveness:

- *Redistribution of the tax burden between firms and sectors*. For some the tax burden will rise, and for others it will fall, depending on their energy intensity and labour use. As stated in the previous section, the loss of market share of the sectors with a higher tax burden competing with foreign firms not subject to the environmental tax will depend on the energy intensity of the sector, their ability to pass costs increases on to prices and their possibility of substituting production technologies and process, but also on the labour intensity of these sectors. Thus, although the overall effect is assumed to be revenue-neutral, this does not necessarily imply that the impact on the trade balance will be neutral.

- *Redistribution of the costs of environmental compliance* (abatement costs): these may be unevenly distributed across firms and sectors. This effect will be greater, the greater the heterogeneity of firms in terms of marginal abatement costs.

- *Reduction in the overall costs of environmental compliance* (total abatement costs), when environmental neutrality is also assumed. Because the environmental tax replaces a pre-existing command-and-control policy, then these total abatement costs fall in aggregate with the introduction of the environmental tax – due to the *static efficiency advantages* of market mechanisms over uniform regulation. They may, nonetheless, rise for some firms.

4.1.3. Double dividend hypothesis and sector competitiveness

As stated in OECD (2001a), the term "double dividend" refers to the possibility that a revenue neutral environmentally related tax shift could generate two possible benefits or dividends. The first dividend is in terms of more effective environmental protection[5] (gains from the static and dynamic efficiency of environmentally related taxes), while the second dividend arises from the reduction in other distortionary taxes. Depending on which marginal tax rates are cut and the specific country considered, the second dividend could generate employment gains, investment gains and/or a more efficient economy, which could counterbalance the competitiveness and equity arguments used against implementing new or higher environmentally related taxes.

Goulder (1995) distinguishes "weak" and "strong" meanings.[6] He defines a "weak" double dividend as the largely uncontroversial claim that using revenues from environmentally related taxes to reduce other tax rates reduces excess burdens, compared with lump-sum return of revenues to the private sector, thus lowering the efficiency cost of the green tax reform. A more controversial claim is that switching the structure of taxation towards a greater revenue contribution from environmentally related taxes would reduce excess burdens. Goulder defines this claim as a "strong" double dividend. Thus, the claim is that a green tax reform does not only improve the environment but also increases non-environmental welfare. If the latter holds, a green tax reform would be "a so-called 'no-regret' option: even if the environmental benefits are in doubt, an environmental tax reform may be desirable" (Bovenberg 1999, p. 421).

The "weak" and "strong" double dividend arguments differ in terms of the comparison that is being made. The "weak" double dividend claim involves a comparison of two cases with *equivalent environmental impact*, but a different use of revenues. It follows directly from the definition of excess burden, which is usually expressed in terms of the impact on economic welfare of a revenue-neutral substitution between a distortionary tax and a lump sum tax. This weak form of the double dividend hypothesis is widely accepted among economists, while the question as to whether the strong form holds, however, depends heavily on the structure of the economy. While a green tax reform is likely to fail to increase non-environmental welfare in economies with functioning labour markets, it *may* succeed in doing so in economies suffering from involuntary unemployment.

The "strong" double dividend claim, that switching the pattern of public revenue-raising towards environmentally related taxes would reduce excess burdens (as well as improve the environment) compares two cases with *different effects on the environment*. It also compares the excess burden of two distortionary taxes (environmentally related taxes and social security contributions, for example), and this is not straightforward, especially in the second-best context where there are existing revenue-raising taxes. Generally, the literature suggests that a double dividend in the strong sense is unlikely to exist, if the starting point has an efficient pattern of revenue-raising taxes. A strong double dividend could arise if initial taxes are suboptimal (although other, non-environmental, tax reforms would also improve welfare, possibly by more), and in a few other special cases (*e.g.* Bovenberg and van der Ploeg, 1994).

From studies of environmental tax reforms in Europe, Hoerner and Bosquet (2001) find that green tax reform packages have tended to reduce the tax burden on labour, primary by cutting non-wage labour costs in the form of social security contributions paid by employers. In their study of simulations of environmental tax reforms, they have found that when the revenues from environmental taxes are used to reduce other distorting taxes, the economic outcome improves both according to employment and GDP. The best employment results is obtained when recycling occurs through cuts in social security contributions, with 85% showing positive employment results, as opposed to *e.g.* 35% positive results for income tax cuts.

The net welfare impact of revenue-neutrally shifting taxes from labour or capital to pollution may be broken down into three components:

1. *a primary welfare* gain that results from the environmental benefits of the reform, net of the reduction in consumer surplus from higher pollution prices;

2. *a revenue-recycling effect*, or efficiency gain, that is positive when the revenue raised by the environmentally related tax is recycled via cuts in distortionary taxation; and

3. *a tax-interaction effect*, which has three components, as per Parry *et al.* (1999). The first component is the efficiency loss from the reduction in labour supply in response to higher pollution prices that reduce real wages. The second and third components are the requirements to raise additional tax revenue on other factors of production (capital in particular) to replace the revenue lost from the reduction in social security contributions and to keep real government spending constant in the face of higher prices. The tax interaction effect has a negative welfare impact, as the burden of the environmentally related tax is partly shifted onto other factors of production, intensifying the efficiency losses of pre-existing tax distortions.

Additionally, dynamic efficiency gains may be realised given ongoing incentives for further emissions reductions to avoid environmentally related taxes on existing emissions levels, through industrial restructuring and the search for cost-efficient abatement technologies and procedures.

Although the "strong" double dividend argument does not create a compelling case for environmentally related taxation, there are a number of significant policy implications which follow from recognition of the "weak" double dividend argument:

1. If revenues from environmentally related taxes are used to make compensating *lump-sum* reductions in tax burdens (*e.g.* for distributional reasons, or to avoid disturbance to sectoral competitiveness), this *foregoes possible efficiency gains*, compared with reductions in marginal tax rates. This generally implies a preference for revenue-raising instruments, compared with those that forego revenues.[7] In choosing between "grandfathered" and "auctioned" tradable permit systems it will be noted that grandfathering is equivalent to a lump-sum return of revenues, and therefore foregoes potential efficiency gains from using the revenues to cut distortionary taxes.

2. *The optimal level of pollution abatement is not independent of environmental policy instrument used* (Lee and Misiolek, 1986). Assuming the rate of the environmentally related tax is below the Laffer maximum-revenue tax rate, the efficient level of pollution abatement will be higher under a revenue-raising instrument than under a non-revenue-raising instrument.

4.1.4. *Competitiveness and different environmental policy instruments*

Any analysis of the competitiveness effects of environmentally related taxes needs to be clear about the comparison being made, and the fact that the effects and optimal tax level are not independent of the instrument used.

Comparing an hypothetical efficient command and control benchmark with an emissions tax achieving the same aggregate level of pollution abatement, the sole difference that would be encountered is the tax burden (distributed between sources according to their residual level of emissions), and the aggregate tax revenues derived. There would be no difference in the cost of environmental compliance (*i.e.* in abatement costs), as compared with the command and control baseline. The additional tax payments by firms may generate costs in terms of a loss in competitiveness; this is perhaps the primary mechanism by which adverse effects of environmentally related taxes on competitiveness are thought to arise. In addition, and perhaps offsetting the effects of this extra tax burden on firms, are the competitiveness consequences of the revenues raised. The issue here is closely related to the "double dividend" literature. If the double dividend literature suggests that employing environmentally related taxes confers a competitiveness benefit (in terms of more efficient revenue raising) on an economy, how far (if at all) does this offset the impact on competitiveness of this environmentally related tax?

Organigram 4.1 – taken from OECD (2003f) – summarises how the set of relevant effects on competitiveness changes in moving from one environmental policy instrument to another. In the diagram, the boxes with bold fonts show various different instruments, while the boxes with normal fonts describe the competitiveness effects which would be experienced in moving in the direction indicated from one instrument to another. It is assumed throughout that all instruments are used to achieve the same overall impact on the environment.

Organigram 4.1. **Different environmental policy instruments in terms of competitiveness impacts**

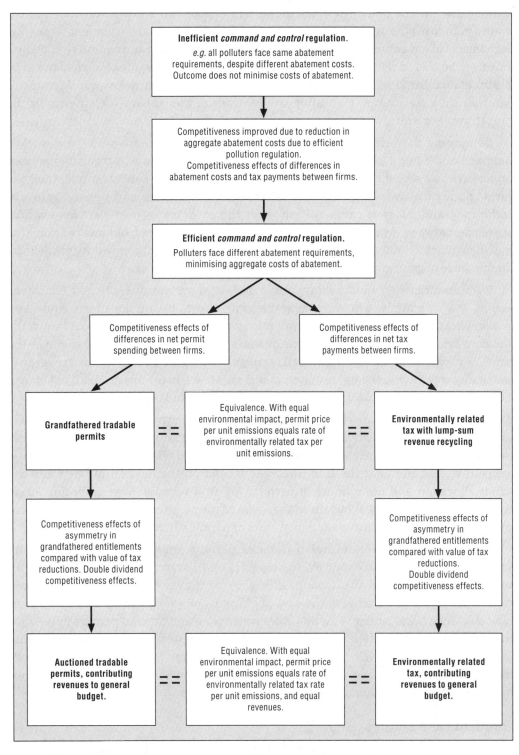

Source: OECD (2003f).

In comparing the competitiveness effects of different environmental policy instruments, a useful benchmark is a hypothetical efficient command and control policy (shown just above the centre), in which the pattern of pollution abatement required across sources minimises the aggregate cost of abatement.[8] The core of the argument for market mechanisms in environmental policy is, of course, that command and control regulation is unlikely to be able to achieve this efficient outcome, because it requires the regulator to be able to obtain full information about firms' abatement costs. Nevertheless, considering this hypothetical case makes it possible to identify various separate elements in the competitiveness effects of different instruments.

Comparing the "efficient" command-and-control benchmark with "real world" command-and-control (shown at the top of Organigram 4.1), in which regulatory policies cannot take full account of the differences in abatement costs between firms, the "real world" policy involves higher aggregate costs of abatement, and hence aggregate inefficiency, and a loss of competitiveness for the economy as a whole. There will be differences between firms: some will incur higher abatement costs, but it is possible that abatement costs for others will be lower. Consequently, differential competitiveness effects – in the sense of gainers and losers among regulated firms – may arise.

With an efficiently functioning permit market, a system of auctioned emissions permits and an emissions tax achieving the same effect on emissions have equivalent fiscal implications. The equilibrium permit price for one unit of emissions would equal the emissions tax per unit, and the payments by firms and the revenues derived would be the same. We can analyse the competitiveness effects of these two instruments interchangeably. Environmental tax instruments are shown in the lower right hand branch of Organigram 4.1, and tradable permits in the lower left-hand branch.

Finally, we can compare the likely competitiveness effects of grandfathered tradable permits, as opposed to auctioned permits. As compared with auctioned permits, no net revenue is derived, and therefore the weak double dividend efficiency gains do not arise. Competitiveness effects at the firm level will depend on the relationship between the permit allocation and the number of permits the firm requires in equilibrium. Firms receiving excess permits will benefit to the value of the surplus permits; firms which are net purchasers of permits incur additional costs of permit purchase.

We can note that grandfathered tradable permits analytically equal emissions taxation where revenues are recycled to taxpayers on the same basis as used for permit grandfathering. Except for the dynamic efficiency incentives the case of grandfathered tradable permits is also similar to the case of efficient command and control. In terms of static efficiency, fiscal burden, etc., this is equivalent to grandfathered permits in the case where the permit allocation exactly corresponds to the efficient pattern of residual pollution (so that no post-allocation permit trades take place). However, under command-and-control firms face no incentive for further innovation in abatement technology, since changes in residual pollution have no financial implications.

4.2. Theoretical case studies

Theoretical case studies of implementing environmentally related taxation are useful for two reasons. First, these simulations help getting a better understanding of the likely impacts of possible cost increases on sectors where these taxes are currently absent. Second, we can use these simulations to test possible measures to reduce industrial relocations when unilaterally implementing environmentally related taxes, for example by using Border Tax Adjustments.

As stated in the previous section, environmentally related taxation in general, and emissions trading schemes (ETS) in particular, could affect the industrial sectors' competitiveness. This section presents two theoretical case studies that evaluate the threat of competitiveness in two sectors: steel and cement manufacturing. While both are high energy-intensive sectors, they differ in the degree of international competition. Steel manufacture has a strong but differentiated international competition; in contrast, cement manufacture has only some degree of international competition.

The theoretical studies presented here indicate modest impact on sector competitiveness – and some sectors will actually gain from such policies, especially if the policies raise revenues that can be used to lower some taxes that cause significant economic distortions.[9] In a similar vein, Carbon Trust (2004) concluded, based on simulations made by OXERA, that:

> "The European Emissions Trading Scheme (EU ETS), properly implemented, will not significantly threaten the competitiveness of most industrial sectors in Europe, including most energy-intensive sectors.
>
> Several sectors have potential to profit from the EU ETS, although there are expected to be winners and losers at an individual company level."

However, the use of economic instruments to significantly reduce greenhouse gas emissions is likely to have negative impacts on the international competitiveness position of some industrial sectors, especially when such instruments are implemented in a non-global manner (unilateral policies). This has, *e.g.* been demonstrated in the case studies of both the steel and the cement sectors; see OECD (2003a) and (2005f).

Finally, before looking more in detail at these studies, it can, however, be useful to keep in mind that the closing down of (otherwise) unprofitable and (also) highly-polluting firms could be the lowest-cost way for society to reach a given environmental target, not withstanding the social implications of doing so. For example, many steel firms have been depending on significant subsidies for the last decades. OECD is working to reduce subsidies to the steel sector in particular[10] and environmentally harmful subsidies in general, see OECD (2003c). Given the present imbalances in the sector, environmentally related taxes may speed up an inevitable restructuring process in the steel industry.

4.2.1. *The steel case study*

This case study presents the simulations results for the steel industry of a widespread use of economic instruments in environmental policy. The objective of the study was to get a better understanding of possible impacts of environmentally related taxes and to assess the foundation for prevailing concerns of unwarranted relocations in the case of unilateral actions.

Why a study on the steel industry?

A simulation case study was undertaken on the steel industry[11] mainly for two reasons: *a)* the energy-intensity of the steel sector and associated CO_2 emissions, and *b)* the competitiveness concern over the business impact of an energy tax.

In terms of environmental significance, the steel industry accounts for about 7% of anthropogenic emissions of greenhouse gas carbon dioxide (CO_2).[12] When mining and transportation of iron ore are included, the share may be as high as 10%. CO_2 emissions from steel production differ significantly between processes, with the "EAF" process involved in recycling scrap releasing less CO_2 per tonne of crude steel than the "BOF" process of the primary integrated steel plants. Primary integrated steel plants transform iron into steel in a Basic Oxygen Furnace (BOF). So-called "mini-mills" recycle scrap into steel in an Electric Arc Furnace (EAF). About 75% of global CO_2 (atmospheric) emissions from steel production are related to the combustion of coal in primary integrated (BOF) steel plants. The remaining emission sources are the use of electric power for scrap melting and the use of natural gas in the production of directly reduced iron. The CO_2 emissions associated with iron and steel production also differ across countries and regions, depending on how much energy is used and the CO_2 intensity of that energy.

Environmentally related taxes and tradable permits will affect the costs of these processes quite differently due to the different input combinations and the resulting differences in emission profiles. More widespread use of economic instruments in the climate policy of OECD countries would increase the costs of steel production. A carbon tax, or an obligation to buy emission permits corresponding to the amount of carbon emitted, would increase the costs of polluting inputs (*e.g.*, coal, oil, natural gas and electricity). Energy costs typically account for 15-20% of the costs of steel production (OECD/IEA, 2000). Hence, a carbon tax may significantly increase the production costs, leading to lower profits, either through lower margins or through a reduction in sales, or both. Reduced profits may in turn lead to closure of firms and/or relocation of activity to countries with less stringent climate polices.

The taxation of the inputs in steel production will not necessarily transform into a one-for-one reduction of profit margins. Part of the tax may be borne by input suppliers (*e.g.*, producers of iron ore and metallurgical coal) through reduced prices on their products. Final consumers, who may accept a higher price of steel products, may carry another part. Moreover, the steel producers themselves may be able to substitute towards less polluting inputs, for instance by increasing the rate of scrap consumption. And finally, there will normally also be possibilities for substitution among different steel producing technologies, at least at the national level. An important part of this study was to assess the possibilities in the steel industry to pass cost increases on to input suppliers and customers, as well as to reduce the cost burden through changes in the input mix or changes in the choice of technology.

Simulations

The study uses a partial equilibrium model of the steel sector to explore short to medium term impacts on the competitiveness of the steel sector of a potential broader use of economic instruments to limit CO_2 emissions. In addition to the steel sector itself, economic instruments are assumed to be applied on fossil fuels used as inputs in electricity generation.

The simulations concentrate on impacts resulting from a hypothetical OECD-wide carbon tax (equivalent to a system of tradable emission permits) levied at USD 25 per tonne of CO_2 emissions. While in practice a CO_2 tax would most likely apply not only to the steel industry but also to industries producing substitutes for steel (*e.g.* aluminium, plastics, wood), the simulation model used does not capture the effects on the steel industry of a broader application of CO_2 taxes.

The carbon tax would induce some substitution from the use of pig iron towards more intensive use of scrap in BOF (Basic Oxygen Furnace) steel making. Scrap prices would then rise, thus weakening the competitiveness of scrap-based EAF (Electric Arc Furnace) steel producers. This in part explains the relatively small difference in net tax burden between BOF and EAF producers compared to the large difference in emissions from these processes. The estimated incidence of the tax can be seen from Figure 4.1.

Figure 4.1. **Tax incidence of levying a CO$_2$-tax in the steel industry**
USD per tonne

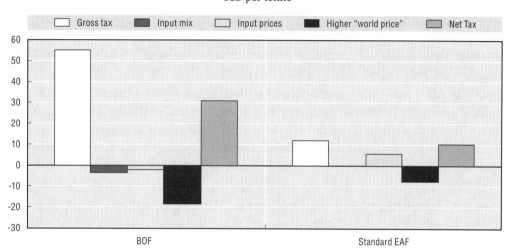

Source: Mæstad (2002).

Because steel demand is relatively price inelastic and because steel is a non-homogenous good, a significant share of the gross tax burden would be carried by the steel consumers through higher prices. The shift of the tax burden over to consumers would be facilitated by the increase in marginal production costs in non-OECD countries as steel producers in this region are pushed closer to their capacity limits.

Suppliers of inputs to the steel industry also carry part of the tax burden of BOF steel producers, but far less than the consumers. For EAF producers, a carbon tax would increase the input costs due to higher scrap prices.

Main simulations results

Table 4.1 summarises the main simulation results. These predict significant reductions in steel production in the OECD area following the imposition of a CO$_2$ tax, in the absence of mitigating measures. Total steel production is estimated to fall by 9% in the short to medium term, with greater reductions (–12%) for the more polluting BOF process plants, compared with the EAF mills (–2%). Steel production outside the OECD, however, would rise by almost 5%, implying a fall in world steel production of roughly 2%.

Despite relatively high emission intensities in non-OECD countries, global emissions from the sector are predicted to decline by over 4.6%, or more than twice the percentage reduction in global steel production, due to a substitution towards cleaner input mix and cleaner production processes in the OECD area. In short, while indicating marked improvement in the environment, the results tend to confirm concerns over the loss of production and accompanying jobs in the steel sector in OECD countries, with

Table 4.1. **Summary of simulation results of a carbon tax levied on the steel sector**

At 25 USD per tonne of CO_2 emissions

	OECD/World steel production	Competitiveness impacts	OECD/World steel CO_2 emissions
OECD-wide carbon tax	↓ 9% / ↓ 2%	Relatively low – Passed on to consumers	↓ 19% / ↓ 4.6%
Unilateral policies	Dramatic cut-backs for integrated steel plants, modest impacts for "mini-mills"	High for integrated steel plants, modest for "mini-mills"	n.a.
Revenue recycling	↓ 1% / ↓ 0.3%	Very low	↓ 10% / ↓ 3%
Border Tax Adjustments	↓ 1% / ↓ 1.8%	None	↓ 10% / ↓ 5.1%

Source: Based on OECD (2003f).

competitiveness impacts differing across producers depending on the process used (BOF: –5.2%; EAF: –0.3%).

On the other hand, unilateral policies by single regions or countries may lead to quite dramatic cut-backs in the production of BOF steel, because unilateralism leaves smaller opportunities to shift the tax burden over to suppliers or customers. For EAF steel producers, the net effect of unilateral policies would not differ much from an OECD-wide approach, because unilateral policies will lead to a smaller increase in scrap prices.[13]

Most of the emissions in the steel industry are related to the consumption of energy. To exempt process-related emissions from the carbon tax would therefore not imply a big relief in the tax burden of BOF steel producers. EAF steel producers would not be directly affected by the exemption for process emissions, but would nevertheless experience a cost increase as BOF steel production would expand and drive up the price of scrap.

If the tax revenues were recycled back to the steel industry as an output subsidy, the decline in OECD steel production would be quite small (< 1%). If the tax refund were uniform across processes, there would, however, be a significant restructuring in the OECD towards the relatively clean process (EAF steel making). This reinforces the fact that different firms within a given sector will be affected in different ways by any use of economic instruments. However, revenue recycling would reduce global emission reductions in the sector from 4.6% to around 3%. In other words, protecting the competitiveness of energy intensive sectors in the OECD area through the recycling of tax revenues to the given sectors are likely to *lower the environmental effectiveness* of the policy as a whole.

So-called "border tax adjustments" (BTA) represent another option to limit sectoral competitiveness impacts of economic instruments. The effect of any border tax adjustments depends crucially on the scope and the design of the adopted scheme. If both import taxes and export subsidies were implemented and were differentiated across steel types, and if the border tax rates were linked to emission levels in non-OECD countries, the decline in OECD steel production stemming from an OECD-wide tax might be as small as 1%. At the same time, the reduction in global emissions (5.1%) would be larger than without border tax adjustments. This is because border tax adjustments keep a higher share of world steel production within the OECD area, thus making more steel producers subject to the OECD-wide carbon tax that was assumed in these simulations. Hence, from an environmental point of view there could, in some cases, be advantages related to the use of border tax adjustments. However, both practical and legal issues related to their implementation need to be solved. See Chapter 5 below for a further discussion.

In the long run, an OECD-wide carbon tax would stimulate investments in new capacity in non-OECD regions. This would reduce the price/cost margin of OECD steel producers even further. However, due to the industry's large sunk costs, there is little reason to believe that the carbon tax in itself would lead to massive closure of firms in the OECD. But given the present imbalances in the steel industry, environmentally related taxes may speed up an inevitable restructuring process in the industry.[14]

Carbon taxes as simulated in OECD (2003f) would seriously hamper new investments in BOF capacity in the OECD area, whereas new investments in EAF steel making would still be profitable. In the long run, a stronger restructuring of the OECD steel industry towards EAF steel making is envisaged.

Limitations of the simulation

The impacts of a carbon tax /quota system will largely depend of the developments of the price of emissions i.e. tax level or quota price. The level of USD 25 per tonne CO_2 used in the analysis is close to the recent level of the quota price in the European Union. The average quota price was for instance approximately EUR 23 per tonne CO_2 in the third quarter of 2005.

Some factors could cause the analysis to overstate the costs of the application of economic instruments:

- Economic instruments are by assumption *not* applied to products that are potential substitutes for steel, like aluminium, plastic, cement, etc. If substitute products also had been included in the analysis, it is likely that impacts on total steel demand of a given price increase would be lower than assumed here.

- Not all potential abatement options in the steel sector are included.

- The model used does not include any endogenous technological change. In the short to medium term – with production capacities largely given – this does not seem like a major restriction. However, in the longer term, considerable technological developments could take place in the sector if carbon taxes or tradable permits were to be applied.

The fact that the simulated price increase is relatively large compared to the value of certain steel products will in general increase the uncertainty of the results.

Finally, impacts on the labour market have not been simulated in this study.

4.2.2. The cement case study

This second case study addresses the competitiveness and leakage issues in the cement sector. It present the results of a spatial international trade model, GEO, merged to a modified version of CENSIM, the world cement model developed by the Institute for Prospective Technological Studies (IPTS). In contrast to the steel case study, the low impact in competitiveness is mainly explained by the ability of the sector to increase operating profits by passing more of their marginal cost increases to prices.

Why a study on the cement sector?

The cement consumption growth over the last decades, the high energy consumption and the very high carbon emissions, from fuel combustion and from the process itself, make the cement sector an important greenhouse gas emitter. The sector's emissions from fuel combustion represented 2.4% of the global carbon emissions in 1994 (IEA, 1999). Adding process emissions, the sector reaches about 5% of the global anthropogenic CO_2 emissions.

In the same time, the cement sector is potentially one of the most impacted by a climate policy: among twelve EU 15 industry sectors, non-metallic minerals – mostly cement – have the second highest direct CO_2 emission/turnover ratio (Quirion and Hourcade, 2004).[15]

Given the recent evolutions of the debate on green house gas (GHG) mitigation, it is clear today that regional rather than global policies will be implemented, at least for a while. Therefore, a distortion of competition may affect countries mitigating GHG emissions through the additional burden of tax policies, emission allowances, etc. Such a distortion may of course have an impact on the competitiveness of GHG-intensive industries. The competitiveness impact and the so-called "carbon leakage" due to this distortion is an argument against non-global mitigation policy or at least in favour of compensations. That is why the quantification of these effects is a priority in the discussion on GHG policies implementation.

The key point to assess competitiveness and carbon leakage impacts of GHG mitigation policies is the representation of international trade. Cement is a relatively homogenous product throughout the world, whose trade is not much disrupted by trade policies or national preferences. Transportation costs[16] and capacity constraints are central to explain international trade patterns. This CENSIM-GEO model deals with market homogeneity, high-transport costs and production capacity shortages more explicitly than in the conventional Armington-based models used in most previous literature.[17] The demand side of the model is based on a commodity intensity curve and a price-elastic demand, while the supply part features seven production technologies, fuel switching, material switching and retrofitting. CENSIM-GEO pays particular attention to the consumption, international trade, energy, emissions, technology dynamics and retrofitting options.

Simulations

A business-as-usual scenario (BaU) from 2000 until 2030 and three climate policy scenarios were built:[18]

- A CO_2 tax or an Emission Trading Scheme (ETS) with auctioned allowances is implemented in the Kyoto Protocol Annex B countries that have ratified it, hereafter labelled "Annex B",[19] assuming a CO_2 price of EUR 15 per tonne.

- The same policy is implemented with Border-Tax Adjustments, i.e. a rebate on cement exports and a taxation of imported cement, is assumed:

  ❖ In the "Complete BTA" scenario (BTA), exported production is completely exempted from the climate policy and imports of cement from the rest of the world are taxed in accordance with the CO_2 intensity of the cement production in the exporting country.

  ❖ In the "WTO BTA" scenario, exports benefit from a rebate corresponding only to the least CO_2 intensive technology available on a large scale, and imports are taxed to the same level.

Main simulations results

Table 4.2 summarises the simulation results.

Business as usual scenario

The business-as-usual (BaU) scenario generated by the CEMSIM-GEO model forecasts an important increase in cement production (2% per year on average until 2030),[20]

Table 4.2. **Summary of simulation results of a carbon price levied on the cement sector**

At 15 Euro per tonne CO2.

	Cement production in 2010, Annex B[1]	Cement production in 2010, ROW[2]	Cement production in 2010, Globally	Cement production in 2030, Globally	CO_2 emissions in 2010, Annex B	CO_2 emissions in 2010, Globally	Main losers, competitiveness impacts
BaU	n.a.	n.a.	n.a.	↑ 2%/yr	↑ 21%	↑ 1.5%/yr (↑ 55%)	n.a.
Annex B carbon tax/ETS	↓ 7.5%	↑	↓	↓	↓ 18% (↓ 22% in 2030)	↓ less than 2% (↑ ROW)	Annex B producers: ↓ Domestic consumption: + Competitiveness: ↓
Annex B carbon tax + BTA[3]	↓ 2%	↓ 1%	↓	↓	↓ 13.5%	↓ 2% (↓ larger than non-BTA) (↓ ROW)	
Annex B carbon tax + WTO BTA[4]	↓ 3% (2010) ↓ (2030)	↑	↓	↓	↓ 15%	↓ (↓ smaller than BTA) (↑ ROW)	Consumers in Annex B: ↑ Cement Price

Note: The values of the three scenarios are compared to the BaU scenario.
1. Countries listed in Annex B of the Kyoto Protocol (expect for the USA and Australia).
2. RoW = rest of the world.
3. BTA = Complete scenario with border tax adjustments: export production exempted and imports taxed.
4. WTO BTA = WTO scenario with border tax adjustments: rebate for exports and imports taxed.
Source: Based on OECD (2005f).

entailing an alarming rise in CO_2 emissions (1.5% per year). The CO_2 efficiency thus rises by 0.5%/yr, thanks to a more intensive use of waste and wood fuels and to the increasing share of modern machines and more energy-efficient technologies.

CO_2 tax implemented in Annex B countries

The implementation of a CO_2 tax, equivalent to a CO_2 Emission Trading Scheme with auctioned allowances, without revenue recycling, in Annex B (except the USA and Australia), at EUR 15 per tonne CO_2 entails a significant decrease in CO_2 emissions in these countries (around 20%). This decrease in emission can be explained by a quicker penetration of energy-efficient technologies, a decrease in the rate of clinker (the CO_2 intensive input) in cement, a quicker switch to low-carbon fuels (gas, waste and wood fuels) and a decrease in cement consumption. However, part of these reductions is compensated by an emissions increase in the non-Annex B countries that export to Annex B. These countries are less CO_2 efficient than Annex B countries and the gap increases with the implementation of the climate policy. Thus, world emissions decrease by around 2% in 2010, 2020 and 2030.[21]

The impact on cement production in theses countries is significant (–7.5% in 2010) because of both a cut in their domestic consumption level and a loss in competitiveness. For the latter reason, production and thus emissions in the rest of the world increase. The corresponding leakage rate is around 25% in 2010 (around 15% after), a result in the upper range of leakage estimates presented in the IPCC third assessment report [5 to 20%, Hourcade and Shukla (2001)].

It must be stressed that the policy tested in this study differs widely from the European emission trading scheme implemented since January 2005, because the study models a tax or *auctioned* allowances whereas in the latter policy, emissions allowances are distributed freely to emitters in a regularly updated quantity. In the EU ETS, in most member States, if an installation is closed, its operator will not receive allowances any more. Conversely, allowances will be issued for free to new installations. Finally, the production level (and possibly the emissions level) will be taken into account in the quantity allocated in the next 5-year periods (Schleich and Betz, 2005). It means that the incentive to reduce both emissions and cement production will be lower than modelled here. As a consequence, these results should not be interpreted as a prediction of what would happen in the case that the European ETS would be implemented in Europe and in the rest of the Kyoto Protocol Annex B.

CO₂ tax implemented in Annex B countries with border tax adjustments

One (efficient) way of preventing carbon leakage and limiting the effects on competitiveness of a non global climate policy is to impose Border Tax Adjustments (BTA). Thus, the last two simulations introduce border tax adjustment by which Annex B countries tax imports of cement from the rest of the world and exempt (at least partially) their exports to the rest of the world from the climate policy are. Two BTA scenarios were tested: a complete BTA and a WTO BTA, which is more likely to be compatible with the current WTO rules.

Under the complete BTA, the loss in production of Annex B countries is limited to 2% instead of 7.5% in 2010. Annex B emissions decrease by 13.5% (31 Mt CO_2, in 2010) thanks to both a production drop and an improvement in the CO_2 efficiency of production. Emissions from the rest of the world also decrease, although very slightly, due to a decrease in their production (around –0.1%). The spillover rate (abatement in non-Annex B over abatement in Annex B) is 6%. Finally, world emissions decrease by 2%, a little more than without BTA. However, compared to business-as-usual, non-Annex B price-competitiveness decreases a little and they lose some market shares, so these countries could claim that this system distorts competition in favour of Annex B countries. Although the policy treats domestic and foreign producers in a similar way (they pay the same cost per tonne of CO_2), it gives a competitive advantage to Annex B producers, who use cleaner production techniques (more energy-efficient technologies and less carbon intensive fuels).[22]

Figure 4.2. **Production in 2010 compared to BaU**

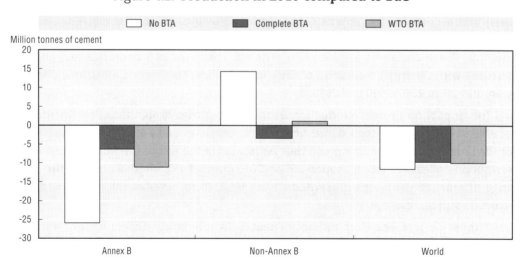

Source: OECD (2005f).

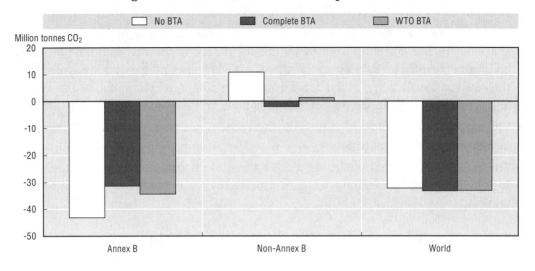

Figure 4.3. **Emissions in 2010 compared to BaU**

Source: OECD (2005f).

Under the WTO BTA scenario, Annex B countries suffer from a slightly higher cost increase than their competitors, which causes a small increase in their imports.[23] However, their exports rise despite this relative variable cost increase, because some of their production capacities become available for exports. Total production actually rises a little in non-Annex B countries, which reduces the rationale for attacking this scenario as distorting competition to the detriment of these countries. Indeed, average cement price in non-Annex B tends to decrease under the pressure of the Annex B, leading to the increase of the consumption. This rise offsets the increase in the net imports of non-Annex B. Emissions from Annex B countries decrease by 15%, more than in the BTA scenario. This is due to the higher drop in their production level. Emissions from non Annex B increase a little compared to BaU, so that the slight spillover observed in the BTA scenario is replaced by a slight leakage, around 4% in 2010. Thus under the WTO BTA scenario leakage is still prevented efficiently, although the reduction in world emissions is a little lower than under the BTA scenario. The main drawback is that the WTO BTA scenario leads, as does the previous one, to a higher increase in cement price and thus hurts consumers in Annex B countries.

Main limitations of the model

These quantitative results should be carefully interpreted, since they depend on assumptions and databases with not always guaranteed reliability. Given the limitations of the model explained below, in general the model overestimates the carbon leakage; however, the main qualitative conclusions seem robust and provide new insights on the much-debated issue of CO_2 leakage and on the ability of BTAs to prevent it.

The main limitations of the model are:

● National demand curves are assumed to be isoelastic. With an isoelastic demand curve, the profit margin remains constant. Impacts on prices and consequently on profits, consumption (the higher is the price, the lower is the demand), production and emissions are significant. However there is no reason, apart from tractability and data availability, to assume that demand elasticities are constant on all the relevant portions of the demand function. This assumption is all the more controversial in a CO_2 constrained scenario which implies a large departure from the baseline conditions.

- The representation of fuel choice is a rigid one, especially waste and wood fuels, the share of which is currently fixed by the model. Indeed in the real world, no technical constraint prevents an increase in the share of these fuels, but few data are available on their cost and availability, which depend largely on other environmental, health-related and agricultural regulations regarding animal flour, waste tyres, etc.

- The model does not allow for a possible increase in transportation costs following the introduction of the climate policy.

- The model assumes that public authorities allow an important use of waste and wood fuels by cement manufacturers, which is criticised by some researchers for public health reasons.

4.3. Political economy lessons from the theoretical case studies

A first lesson that we can draw from the theoretical case studies is that *economic instruments will not affect different firms within a given sector in the same way*. This is explained mainly by the different input combinations and the resulting differences in emission profiles.[24] An example can be found in the steel case study: when tax revenues were recycled back to the steel industry as an output subsidy, the decline in OECD steel production was estimated to be quite small (< 1%). If the tax refund were uniform across processes, there would, however, be a significant restructuring in the OECD towards the relatively clean process, *i.e.* EAF steel making.

However, an important point to note here is that *revenue recycling would reduce global emission reductions* in the steel sector (from 4.6% to around 3%). In other words, a second lesson is that protecting the competitiveness of energy intensive sectors in the OECD area through the recycling of tax revenues to the given sectors are likely to lower the environmental effectiveness of the policy as a whole.

A third lesson is the *importance of taking into account possible adjustments in related markets* when considering the impacts of a given policy on a particular sector. A part of any initial burden placed on a sector is likely to be shifted backwards to input suppliers and forward to the customers.[25] In the steel case, this is *e.g.* illustrated by the (somewhat crudely) estimated impacts on scrap metal prices, and the increase in steel prices. Because steel demand is relatively price inelastic and because steel is a non-homogenous good, a significant share of the gross tax burden would be carried by the steel consumers. The shift of the tax burden over to consumers would be facilitated by the increase in marginal production costs in non-OECD countries as steel producers in this region are pushed closer to their capacity limits. Suppliers of inputs to the steel industry also carry part of the burden of BOF steel producers, but far less than the consumers. For EAF producers, a carbon tax could increase the input costs due to higher scrap prices.

A fourth lesson is that, in spite of *some element of* "carbon leakage" even when policies to combat climate change are put in place on a relatively broad front, *significant global reductions* in carbon emissions – compared to a reference scenario – can be achieved.[26] Thus, an OECD-wide tax would reduce OECD emissions of CO_2 from the steel industry by 19%. Despite relatively high emission intensities in non-OECD countries, global emissions from the sector would decline by 4.6%, *i.e.* more than twice the reduction in global steel production. This is due to substitution towards a cleaner input mix and cleaner processes in the OECD area. A similar finding is made in the cement sector case study. In a scenario with emission trading in Annex B except USA, even if cement production outside this area increases in response to the policies put in place in most of the OECD-related regions, global carbon emissions in clinker manufacturing are estimated to decrease.

As a fifth lesson we note that – while the importance varies between different industries – *the larger the group of countries that put similar policies in place the more limited the impacts on sectoral competitiveness*. The steel case study shows a low impact on the competitiveness of the sector when an OECD-wide carbon tax is implemented because of the ability to pass the tax to the consumers. In contrast, unilateral actions reduce these opportunities to shift the tax burden over to suppliers and customers and, therefore, the impact on competitiveness turns high for integrated steel plants and modest for "mini-mills".

Finally, from an environmental point of view there could, in some cases, be advantages related to the use of border tax adjustments, as simulation results shown in both case studies. However, the effect of any border tax adjustments depends crucially on the scope and the design of the adopted scheme and both practical and legal issues related to their implementation need to be solved.

Notes

1. In addition, it is a relevant issue whether or not given environmental targets – or the lack of such targets – represent a reasonable balance between the benefits and costs of environmental improvements. The choice of policy instruments can, however, affect the efficiency at which a given target it reached – and affect the rate of new technology developments, which can be important for the cost to society of reaching given policy targets in the longer term. Economic instruments, like taxes or tradable permits, can help in achieving a given target at the lowest possible cost to society as a whole, both in the short term, as they can equalise marginal abatement costs between polluters (obtain static efficiency), and in the long term, as they provide a continuous incentive for further technology development.

2. It is relatively uncontroversial that some firms can benefit from undertaking more rigorous environmental protection. A more controversial claim is that made by Porter and van der Linde (1995) that whole economies can enhance their competitiveness through more stringent environmental standards.

3. Even if there are reasons to provide (temporary) relief to those that lose out from the change, it is by no means given that this best can be done through some modification to the environmental policy.

4. The smaller and more open the country, the more exposed will firms to be externally given world market prices and conditions, and the smaller will be the scope for shifting tax burdens on to customers or suppliers through price incidences.

5. Environmentally related taxation can also generate supplementary environmental benefits, for example a revenue-neutral CO_2 tax, would create incentives to burn less fossil fuels, which in turn could reduce other emissions associated with fossil fuels, for example SO_2 emissions. On the other hand, some measures to reduce CO_2 emissions *could* lead to increased NO_x emissions – and vice versa.

6. Goulder (1995) surveys the early literature; more recent surveys on the double dividend hypothesis are Bovenberg (1995, 1998, 1999), Bosello, Carraro and Galeotti (2001) and Schöb (2003).

7. After all, if the environmental externality is not addressed by the tax, it has to be addressed by some other form of regulation and there are no guarantees that these other forms of regulation will not have similar counterintuitive effects (Sterner, 2003). After all, all regulations raise prices initially.

8. It is convenient to assume that the context is one where emissions from different sources are equally damaging to the environment. Adding the possibility that pollution damage might vary between sources complicates the analysis, without introducing any useful additional insights on competitiveness effects.

9. Even without any revenue recycling, *some* sectors are likely to improve their international competitiveness position as a result of domestic adjustment in prices and wages over the longer term.

10. See *www.oecd.org/document/5/0,2340,en_2649_34221_32362885_1_1_1_1,00.html*.

11. Report prepared by Dr. Ottar Mæstad of the Institute for Research in Economics and Business Administration in Bergen (Norway) for the OECD's Joint Meetings of Tax and Environmental Experts. The reported results are largely based on simulation experiments with the Steel Industry Model (SIM). As there are uncertainties attached both to parameter values and the functional forms used in the model, the exact figures that come out of such numerical analyses should be treated with caution. Emphasis should instead be put on the qualitative insights.

12. See Ecofys (2000).

13. Simulations made by The Carbon Trust (2004) on the effect of the imposition of the European ETS also show that unilateral policies (EU ETS) will imply a loss of competitiveness for the EU companies in the long term if other major producers did not face carbon constraints. It also concludes that the steel industry should not lose out as a result of the EU ETS out to 2012, although this result depends on the extent to which EU companies can pass cost-increases to consumers.

14. This result is also found in a study of the German steel industry, Lutz *et al.* (2002), where the authors conclude that a CO_2 tax of the magnitude considered in OECD (2003f) will induce significant technological development towards cleaner production process. They also argue that the burden of CO_2 taxes is likely to be reduced in the long run, making closure of firms less likely than it otherwise would be. This can be possible by both innovations within existing technological paradigms and possibilities of further developing of new technological paradigms with lower emission rates.

15. Only electricity generation has a higher ratio, but this sector is largely sheltered from international competition by transmission losses.

16. According to OECD (2005f), whereas average cement price in Europe was around EUR 70 per tonne of cement in 2004, the sea transportation cost of a tonne of cement between Greece and Spain was around EUR 22, and the road transportation cost in France around EUR 8 per tonne for 100 km. Therefore, cement generally does not travel more than 200 km by road between the plant and the consumer.

17. The much-used Armington specification assumes that goods of the same kind produced by different countries are not perfect substitutes. This imperfect substitution, while reflected in only one parameter – the Armington substitution elasticity or a parameter with an equivalent meaning – has various grounds: products are not homogenous throughout the world, consumers have national preferences, trade policies and transportation costs constitute barriers to trade. In some models, transportation costs between countries are taken into account in prices, using a fixed transportation cost or real distances between national capitals. However, the countries are still treated as dimensionless points.

18. The CEMSIM-GEO model includes 47 producing countries aggregated in 12 regions: Europe (EU25, Bulgaria, Romania and the rest of western Europe), R&U (Russia and Ukraine), Japan, Canada, the USA, RJAN (Rest of Japan, Australia and New Zealand), TRR (Turkey, Rest of CIS and Rest of Central and Eastern Europe), Latin America, India, China, Rest of Asia and A&ME (Africa and Middle-East).

19. The United States and Australia are not included in the Annex B area in these simulations. Since New Zealand is merged with Australia in the model data set, it has been assumed that it does not implement the climate policy, although is has ratified the Kyoto Protocol.

20. At the world level, cement consumption is estimated to increase from 1 630 Mt in 2000 to 2 900 Mt in 2030, corresponding to an annual 2% growth rate.

21. This amounts to a global emission reduction of 32 Mt CO_2 in 2010, 39 Mt CO_2 in 2020 and 34 Mt CO_2 in 2030.

22. On the cement markets inside Annex B, every country's variable production cost increases by an emission cost, which varies from country to country in accordance to the CO_2 intensity of its production. Under the Complete BTA, Annex B variable production cost in Annex B markets increases more than in non Annex B countries mainly because the former are in general more carbon efficient than the others. Indeed, not only do they use more energy efficient technologies and less carbon intensive fuels already in BaU, but the climate policy also leads them to reduce their CO_2 emissions per tonne of cement (especially by decreasing their clinker rates), while non-Annex B countries do not.

23. Under the WTO BTA scenario, the emission cost for non-Annex B countries in Annex B markets is limited by the one of the less carbon intensive technologies.

24. Figure 3.3 in Chapter 3 provides an *ex post*, "real world" illustration of this point.

25. It is very important to keep this point in mind these days, when many sectors in most OECD countries are asking for compensation due to the high crude oil prices. The current crude price increase is an "exogenous shock" that affects (almost) all firms in the relevant sectors across the World, and it seems very likely that they will *gradually* be able to shift a significant share of the cost increase on to their customers or their (non-energy) input suppliers.

26. The word "significant" is of course relative. The emission reductions obtained in the simulations discussed here are small compared to what would be needed to fulfil the long-term objectives of the UN Framework Convention on Climate Change.

ISBN 92-064-02552-9
The Political Economy of Environmentally Related Taxes
© OECD 2006

Chapter 5

The Sectoral Competitiveness Issue – Border Tax Adjustments

5.1. Introduction and background

This chapter examines the extent to which the GATT/WTO rules permit the use of border tax adjustments (BTAs) to address potential competitiveness issues arising from the implementation of environmentally related taxes. It addresses the extent to which inputs which are embedded or used in the production of goods, such as energy, are eligible for border tax adjustments. Lessons from the US's experience with environmental excise taxes will provide the contemporary context for the analysis with a view to locating their relevance to the treatment of unincorporated inputs used in the production process.

Concerns about the impacts that human activity are having on the earth's atmosphere in the late 1970's to early 1980's led to concerted efforts by governments around the world to examine ways of addressing the challenges. The forums that provided the platform for the deliberation of these issues were the 1992 "Rio Earth Summit" which resulted in the signing of the United Nations Framework Convention on Climate Change in 1992, and the ensuing negotiations which culminated in the ratification and coming into force of the Kyoto Protocol in 2005.

In the wake of these developments, some OECD countries have resorted to the use of economic instruments such as environmentally related taxes and charges in conjunction with regulatory measures to address the challenges posed by the fallout from global warming. This has resulted in a marked shift from the traditional use of taxes on energy and fuel for fiscal purposes towards their use for environmental purposes. To achieve this objective, energy taxes have been designed to raise the price of energy to incorporate the cost of environmental externalities associated with their use. This adaptation has been justified on the basis of a principle saying that the polluter should pay, where the objective is to reduce greenhouse gas emissions by curtailing consumption, encouraging more efficient fuel use, shifting consumption from fossil fuels to other energy sources, and penalizing energy intensive industries.[1] Because a tax conveys the same message to all emitters, those who can reduce emissions at a low cost will do so. Different schemes of carbon/energy taxes have been introduced in a number of OECD countries. As described in Chapter 2, there are currently 150 environmentally related taxes in OECD countries that are levied on energy products.

5.1.1. Competitiveness issues

The unilateral introduction of energy taxes in some OECD countries has raised concerns about their price effects not only on energy prices but on those products whose production gives rise to a large amount of emissions. These concerns have related to the international competitiveness of domestic products, particularly those in the energy-intensive sectors. From the perspective of domestic producers, the unilateral introduction of these taxes have been perceived as subjecting them to a double jeopardy, in the sense that they have a "penalty" effect on them as a result of their having to compete with imports that may not have been subject to similar taxes, whilst at the same time they have had to compete with similar untaxed products on the international market.

The imposition of taxes on energy has resulted in significant price differentials in energy between countries with energy tax regimes and other countries. Biermann and Brohm (2003) estimate the price of heavy fuel for industry in the US to be one fifth lower than an average of nine other OECD countries that have energy taxes.[2] Electricity prices for industry in the US are lower by a third. EU taxes on gasoline have also been estimated to range from 66% to 81% of the end use price compared to around 30% in the United States (WTO 2002). Figure 2.6 above and the accompanying text also provides a comparison of the differential tax rates for petrol and diesel in OECD countries.

The question of the impact that taxes have on trade and investment flows has been highly polarised in tax literature. An examination of the issues would be beyond the scope of this chapter. In considering the competitiveness effects of energy taxes, it is important to distinguish between competitiveness impacts at a national and at a sector or firm level. Most economic studies have not been able to establish a direct causal link between environmentally related taxes and competitiveness at national level – as they have been limited by the fact that most carbon/energy tax regimes provide either a full or partial exemption for heavy industries and export industries. Furthermore, the effective tax rates under current energy taxes are generally believed to be too low to achieve the reduction thresholds of the Kyoto Protocol [Biermann and Brohm, (2003); OECD, (1996)]. These considerations have made it difficult, if not impossible to measure the competitiveness impacts at the national level. These notwithstanding, it is now established that the use of economic instruments to significantly reduce greenhouse gas emissions is likely to have negative impacts on the international competitiveness position of some industrial sectors, especially when such instruments are implemented in a non-global manner. This has for example been demonstrated in recent OECD case studies of the steel and the cement sectors, cf. the discussion in Chapter 4 above.

International competitiveness concerns have been responsible for the scrapping of proposals to introduce the 1993 BTU Tax legislation in the US, the "Greenhouse Levy" in Australia in 1994, and the EU Council's Directive to establish a common EU framework on energy taxation in 2003.

5.1.2. Policy instruments for addressing competitiveness concerns

Governments have used two policy instruments to address the international competitiveness issues associated with the introduction of environmentally related taxes and charges.

The first has been to grant exemptions to the most energy-intensive industries. This has been the most common approach used in Europe for high energy taxes. Most OECD countries implement a range of strategies to relieve or exempt polluting or energy intensive industries from environmental taxes.[3] At the introduction of the Dutch energy tax large-scale users of electricity were totally exempted. Later brackets with regressive rates were introduced. Exemption is given for the use of electricity above 10 000 000 kWh if an enterprise has concluded an agreement with the government on energy-efficiency. In many cases, energy products used mainly by heavy industry are exempted from tax. Most countries do not tax coal at all, while the few countries that have taxes on these products grant very significant exemptions. In other cases, reduced tax rates combined with generous rebates are applied to industry in respect of carbon or energy taxes, as in the case of Denmark Germany, Sweden and United Kingdom.

The second policy instrument is to provide for border tax adjustments (BTA) where companies have environmentally related taxes rebated to them upon export and have domestic environmental taxes added to imports. When implemented in accordance with the "destination principle", it enables each country to tax its domestic industries for internal purposes whilst preserving its competitiveness internationally. It also allows its exports to compete in untaxed markets abroad, whilst ensuring their competitive advantages domestically by taxing imports up to the same level.

All countries which levy domestic taxes on fossil fuels for fiscal purposes do apply a border tax adjustment equal to the domestic tax when importing such fuels (OECD, 1997). BTAs have also been used in the US in two important instances of environmental excise taxes: the Superfund Chemical excises (Superfund Tax) and the Ozone Depleting Chemicals Tax. No border tax adjustment schemes currently exist for taxes on energy inputs used in the production of final goods (Biermann and Brohm, 2003). They also state that current energy price differentials between parties to the Kyoto Protocol and non-parties do not justify the use of these measures at this stage.[4]

The imposition of taxes on imports or exemption/rebating of taxes on exports are obviously barriers to trade, so they do come within the scope of the multilateral trading system, as they do raise trade law concerns. If a government generally imposes an energy tax but then exempts particular industries, such exemption could be treated as a specific subsidy that is actionable under the 1994 Agreement on Subsidies and Countervailing Measures (1994 ASCM). Similarly, if an exemption is targeted at export-oriented industries, it could be perceived as a prohibited export subsidy under the 1994 ASCM.

From a fiscal perspective, while exemptions for energy-intensive or export-oriented industries can eliminate competitive disadvantages, they do create significant problems. Extensive exemptions eliminate or reduce the intended tax incentive of developing more carbon-efficient production processes or shifting to carbon-neutral energy sources, such as renewable energy. Further, such measures dilute the environmental impact of an energy tax by undermining the full internalisation of the external costs by the producer, see Biermann and Brohm (2003).

However, the application of BTAs to energy taxes under the GATT/WTO rules is clouded with uncertainty. This is due to the fact that the treatment of inputs (such as energy) which have not been physically incorporated into a product has never been formally considered by a GATT/WTO dispute panel.

5.2. Border Tax Adjustments: their historical and contemporary context

The use of border adjustments for excise taxes has a long history. The first US excise tax, imposed on distilled spirits in 1791, was adjusted at the border by taxing imports and exempting re-exports, see Hufbauer (1993). In the late 19th century, rules for the use of border tax adjustments were included in intergovernmental agreements to prevent the protectionist use of this instrument. Later, this issue became relevant in the negotiation of the GATT. However, it was not until the late 1960s that it became an issue which caught the attention of policy-makers. At that time the European Community required its members to replace their national "turnover taxes", or sales taxes, with a value added tax (VAT). By 1970, all members of the European Community had to adjust their VAT at the border according to the destination principle. It was during this period (the late 1960s) when some countries began to harmonise indirect taxes that discussions about border tax adjustment emerged within the OECD trade committee and the GATT.

Against this background, the GATT set up a Working Party on Border Tax Adjustments in 1968 to examine various issues that had emerged from these developments. In its final report, the Working Party (in paragraph 4) *inter alia*, adopted the following definition of BTAs as applied by the OECD:

"… any fiscal measures which put into effect, in whole or in part, *the destination principle* (*i.e.* which enable exported products to be relieved of some or all of the tax charged in the exporting country in respect of similar domestic products sold to consumers on the home market and which enable imports sold to consumers to be charged in the importing country in respect of similar domestic products)."

This definition was later confirmed by the WTO Committee on Trade and Environment, see WTO (1997b).

The US has had a long history of using border tax adjustments in its excise legislation as a means of maintaining the competitiveness of its export sector, Hoerner (1997).

The responses that governments have adopted to address the challenges of climate change has rekindled the interest in the use of this policy instrument. A 1996 Research Panel Report prepared for the Japanese Environment Agency suggested the possibility of using border tax adjustments "for products exchanged in the international market when dealing with countries that do not take similar economic measures to protect the environment", see Government of Japan (1996). This report has been the subject of ongoing discussions between the Environment Agency and Japanese industry, see Japanese Environment Agency (1997).

5.3. The GATT regulatory framework governing BTAs

5.3.1. Non- discrimination as the foundation for competition

GATT provisions (and the Articles I and III in particular), are premised on outlawing discrimination in international trade. To that extent, they have been inherently designed to foster competition.

Article I of GATT, dealing with the *most favoured nation clause*, bans discrimination based on the origin of goods with respect to customs matters, internal taxes and internal sales regulations. It does so by requiring *contracting parties* (*i.e.* members) to extend the favourable treatment of a foreign product to *like* products from all other contracting parties.[5]

Article III of GATT, which is like the flipside to a coin to Article I, secures the *national treatment* of products of the contracting parties that are imported into the territory of other contracting parties with respect to internal taxation and regulations. Effectively, this means that members must not discriminate between domestic and foreign commodities once they have crossed the border. Foreign products shall be given national treatment, *i.e.*, the same treatment as national products.

5.3.2. Interaction between the provisions on non-discrimination with the BTA rules

In the context of applying the rules to BTAs, Article II:2 (a) expressly indicates that it operates in conjunction with Article III:2. Article II:2(a) of GATT enables the contracting parties to impose internal taxes on imported products, by providing that nothing in Article II (on schedules of concessions) shall prevent parties from imposing on the importation of any product

"… a charge equivalent to an internal tax imposed consistently with the provisions of paragraph 2 of Article III in respect of the like domestic product or in respect of an article from which the imported product has been manufactured in whole or in part."

Under Article II therefore, contracting states are legally obliged to refrain from raising border taxes in the form of tariffs above the specified rates agreed in GATT negotiations and incorporated into its schedule of concessions. The tariff rates so agreed are known as "bound" rates. Their purpose is to provide greater commercial certainty through the establishment of a ceiling on tariffs that cannot be breached without an offer of compensation to affected trading partners. It also permits the imposition of domestic taxes and charges, provided that it is done in accordance with the *national treatment requirement* in Article III. Article III:2 sets the framework that deals with the ability of the contracting parties to impose internal taxes on the imported products from other contracting parties.

In addition, an interpretive note to Article XVI of GATT provides that the exemption or remission of internal taxes on exported products is not a subsidy in terms of Article XVI of GATT. The combined effect of Articles II, III, and the interpretive note to Article XVI of GATT have been interpreted as allowing BTAs according to the destination principle, see GATT (1970).

5.3.3. Internal taxes[6]

The texts of these articles raise the question of the categories of internal taxes that are eligible for Border Tax Adjustment relief under the GATT rules. This issue was examined by the GATT Working Party on Border Tax Adjustments [See GATT (1970)] which concluded that there was "a convergence of views to the effect that taxes directly levied on products (*i.e.*, indirect taxes) were eligible for adjustment". Examples of such taxes include specific excise duties, sales taxes, and value added taxes. BTAs for these taxes are commonplace in practice.

The Working Party further concluded that there was also "a convergence of views to the effect that certain taxes that were not directly levied on products (but on the producer, *i.e.* direct taxes) were not eligible for tax adjustment." Examples include social security charges and payroll taxes. This list was further expanded upon in the 1994 Agreement on Subsidies and Countervailing Measures which defined direct taxes to include taxes on wages, profits, interests, rents, royalties, and all other forms of income, and taxes on the ownership of real property.[7]

The differential treatment of direct and indirect taxes has been justified on the basis that indirect taxes are shifted forward by the taxpayer so as to be reflected in the price of the product, and that direct taxes are not (OECD, 1994). If indirect taxes are shifted forward and direct taxes are not, then Border Tax Adjustments act to preserve competitive equality in international trade, neither granting subsidies or incentives to exports nor disadvantaging imports relative to domestic production. Although there are few records of discussion of these points during the drafting of GATT, substantially similar provisions had been used in commercial treaties for some time. The GATT Working Party on BTAs also reached agreement that these provisions reflected the general practice at the time the GATT was signed.

5.4. Border Tax Adjustments on final products

5.4.1. Background

The GATT rules have been interpreted as allowing BTAs with respect to taxes on inputs which are "physically incorporated" into the final product. Thus, taxes on inputs such as the main chemical constituents of chemical products and the iron and raw materials used in steel production could be applied to imports and rebated on exports.

5.4.2. Border Tax Adjustments on imported goods

There are two separate bases under which levies are permissible in dealing with the imposition of border tax adjustments on imports under Article III:2. The first involves items that are categorised as "*like products*"; and the second deals with products that are "*directly competitive or substitutable*".

The central requirement for the imposition of indirect taxes on imports that are eligible for BTAs is that they must comply with the *national treatment principle* in Article III:2 of GATT. Article III:2 of GATT provides:

> "The products of the territory of any contracting party imported into the territory of any other contracting party shall not be subject, directly or indirectly, to internal taxes or other internal charges of any kind in excess of those applied, directly or indirectly, to like domestic products *(First limb)*. Moreover, no contracting party shall otherwise apply internal taxes or other internal charges to imported or domestic products in a manner contrary to the principles set forth in paragraph 1" *(Second limb)*. *[Emphasis added.]*

5.4.2.1. Article III: 2: First Limb – "Like products"

The GATT's "*like product*" provisions impose an obligation on members to treat foreign products from other member countries as like national products. The provisions were drawn up to outlaw discrimination on the basis of the national origin of the product in question.

The application of this limb of Article III:2 is dependent on two criteria. First, a determination as to whether the domestic and imported products are *like*. The term "*like product*" has not been defined anywhere in the GATT. The GATT Working Party on BTAs reported that it had been unable to develop a definition and concluded that problems arising from the interpretation of the term should be examined on a case-by-case basis. The Working Party suggested that relevant factors that could be taken into account in determining whether products were *like* included "the product's end uses in a given market; consumers' tastes and habits, which change from country to country; the product's properties, nature and quality", see GATT (1970).

The tests for determining whether an imported product is a *like* domestic product, have evolved over time.[8] In the *Tuna-Dolphin* Case, the Panel in determining the meaning of the term *like*, in the context of Article III:4 concluded that differences in the production processes of a product could not be taken into account in determining the likeness of the domestic and the imported product so long as the production process did not affect the product as such, see GATT (1991).[9] In a later decision, the *Shrimp-Turtle* Case however, the WTO Appellate Body suggested that process-based trade measures might still be acceptable under Article XX if applied in a non-discriminatory way, see WTO (1998b). The decision is significant, as it loosened the ban on trade measures based on process standards formulated in the *Tuna-Dolphin* decision.

With respect to the term "like products" in Article III:2, the *US Alcohol Case* dealt with identical products with different origins and ingredients. In determining whether the products were "like", the Panel stated that account had to be taken of the purpose of Article III, namely, the prevention of contracting parties from using their regulatory powers for protectionist purposes. The Appellate Body in the *Japan Alcohol* case determined that the notion of "*like product*" did not mean that the imported and the domestic products had to be identical. It likened the concept to the metaphor of the accordion which could stretch and squeeze contingent upon the context of the article in which it appeared, *cf.* WTO (1996c).

Once the imported and domestic products are determined to be like, the second criterion that has to be satisfied is that the taxes and charges applied to imported products "must not be in excess of those applied to *like* domestic products". This requirement has been construed to preclude the imposition of taxes at a level in excess of domestic products even if the differential is very small, see GATT (1987).

5.4.2.2. *Article III: 2: Second Limb – Directly Competitive or Substitutable Products*

The second limb of Article III:2 of GATT, when read together with Article III:1 of GATT and the interpretive note to Article III:2, require that imports should not be taxed dissimilarly from "directly competitive or substitutable" domestic products in a manner intended "to afford protection to domestic production". Unlike the first limb of Article III:2, the second limb is based on a much broader criterion, namely the protective nature of the of internal tax system. The second limb will therefore only apply if the imported and the domestic products are not *like* products. In determining whether imported and domestic products are alike, the applicable test is whether they satisfy the requirements set out in the first limb of Article III:2 of GATT.

Two major issues arise in applying the second limb. The first relates to the determination of "competitiveness or substitutability" of products. The competitiveness or direct substitutability of products is determined by reference to the criterion of whether the products have common uses, as demonstrated by their functional substitutability, intended use and price [See WTO (1996c) and Schoenbaum (1992)]. Provided two products perform the same function for consumers, important social or environmental differences are seen as irrelevant. Products will be distinguishable when consumers in the relevant country perceive there to be a significant distinction, thus giving the second limb of Article III:2 an "element of dynamism" that is absent from the "like product" language, see Fauchald (1998).

The imposition of differential taxes on functionally equivalent products, based on either physical characteristics or processes and production methods, would be subject to the second limb of Article III: 2. It is arguable that where a buyer's decision is influenced by environmental factors, the method of production also becomes an important consideration for assessing a product's relevant characteristics. Such differentiation is exampled in some consumers' preferences for "organic" fruits and vegetables, "non-battery hen" eggs, and dolphin-safe tuna [See Cameron (1993); Hoerner and Muller (1996); Goh, 2004].

The second issue is whether the imposition of a heavier tax burden on imported products relative to domestic products could be construed as being applied in a manner as to afford protection to domestic products? This question requires a separate determination of the protective intent. The protective determination is ascertained by objectively considering its design, underlying criteria, architecture and structure, see WTO (1996c). The geographical distribution of discriminatory effects will provide guidance. The mere fact that the burden of the tax is primarily borne by imported products will not reveal a protectionist intent if the country of origin is merely coincidental.[10] Where the criteria or characteristics that attract the tax are inherently limited to products from specific countries, such as taxes on tropical timber, protective application will be easier to establish.[11] The magnitude of the tax differential may also, according to WTO (1996c), provide evidence of protection. From the perspective of using border tax adjustments as a means of overcoming competitiveness concerns in relation to the introduction of environmentally related taxes, this could raise possible complexities.

5.4.3. Border Tax Adjustments on exports

The competitiveness impacts of environmentally related taxes and charges on domestic industry can also be abated through the remission of internal taxes or charges on products destined for export. To the extent that consumption externalities do not contribute to global environmental harm, relieving exports of the domestic tax burden reflects a more appropriate allocation, since it is presumed that these will be dealt with in the consumer country. Where consumption externalities are transboundary, it might be more appropriate to limit the scope for the remission of environmentally related taxes.

The application of the destination principle to exports raises the question of whether export rebates provided in those circumstances would amount to a prohibited or countervailable export subsidy under the GATT rules. This issue stems from the fact that Article XVI:4 of GATT prohibits direct or indirect subsidies for products that result:

> "… in the sale of such product for export at a price lower than the comparable price charged for the like product to buyers in the domestic market,"

Articles VI:3 and VI:6(a) of the GATT allows a party to impose countervailing duties on the imported product equivalent to the subsidy granted by the exporter.

The interpretative note to Article XVI of GATT 1994 sets out three conditions which govern the treatment of border tax adjustments for exported products. First, only those internal taxes which are actually levied on domestic products are eligible for exemption or remission upon export. Second, the exemption or remission of these internal taxes must not exceed the amounts which were levied on the domestic products. Third, the exported and domestic products must be alike.

The need to impose stringent conditions for the treatment of BTAs on exports is underscored by the fact that they can be used as a disguise for primarily removing competitive disadvantages for domestic industry.

Article VI:4 of GATT 1994 precludes the application of countervailing duties where an exported product has been exempted from domestic taxes. These principles have also been mirrored in the 1994 ASCM. The 1994 ASCM also dealt with the availability of export rebates to taxes on a particular product or on inputs that have physically been incorporated into the product.

5.5. BTAs on inputs not physically incorporated into products

5.5.1. Placing the issue in context

Whilst the GATT allows border adjustments for internal taxes on final goods, when it comes to the application of border tax adjustments for processes or on inputs which are not physically incorporated into the final product, the issues are much more complex.

A concern with the remission of taxes on embodied inputs for BTA purposes is the potential difficulties that it poses from an administering and enforcement point of view. The complexities of the underlying issues in this area present a potential minefield from a regulatory perspective. For example, measurement and administration problems can be encountered when providing relief to exports of steel in respect of the amount of carbon tax embedded in the pre-adjustment export price. An estimation could give rise to inequitable and distorting outcomes if different production technologies are used by different producers, in different countries over time, implying a varying carbon content. Thus the administrative and compliance costs with establishing the correct BTA in such cases can be quite significant.

The 1970 GATT Working Party considered this issue but was unable to reach a conclusion. It noted that "there was a divergence of views with regard to the eligibility of certain categories of tax", such as the *taxes occultes*, which encompass consumption taxes on capital equipment, auxiliary materials and services used in the transportation and production of other taxable goods, as well as taxes on advertising, energy, machinery and transport. The Working Party did not investigate the matter; it felt that "while this area of taxation was unclear, its importance – as indicated by the scarcity of complaints reported in connection with adjustments of *taxes occultes* – was not such as to justify further examination".

5.5.2. The state of the debate on the issue

Expert commentary is also divided on the potential applicability of the GATT rules in this context.

Brack *et al.* (2000) state that the exemption or remission of indirect taxes on manufacturing inputs that are embodied without being physically incorporated into the traded good (*taxes occultes*), where exports are concerned is prohibited under the 1994 Agreement on Subsidies and Countervailing Measures, and its predecessor, the 1979 Subsidies Code.

Footnote 61 to Annex II of the 1994 ASCM adds the definition that "inputs consumed in the production process are inputs physically incorporated, energy, fuels and oils used in the production process and catalysts which are consumed in the course of their use to obtain the exported product". In effect, taxes on the items listed in footnote 61 – including "energy, fuels, and oils used in the production process" can be adjustable at the border, at least for exports. Unlike the situation for imports, this would appear to permit BTAs for energy or carbon taxes based on processes.

Brack *et al.* (2000), however, point out that the entire Annex (*i.e.* Annex II) relates to "*prior-stage cumulative indirect taxes*", which are further defined in the Agreement as follows:

> "*Prior stage*" *indirect taxes* are those levied on goods or services used directly or indirectly in making the product.

"Cumulative" indirect taxes are "multi-stage taxes levied where there is no mechanism for subsequent crediting of the tax if the goods or services subject to the tax at one stage of production are used in succeeding stages of production."

In their view, the earlier term used, *taxes occultes* – such as excise duties – are arguably not "cumulative" taxes because the carbon or energy is taxed only once – at the point of its inclusion in the production process. By way of an example, even though energy is used in every stage of the manufacturing process, each unit of fuel is only taxed once. The archetypal PSCI tax is a cascade tax, which is an *ad valorem* tax on all transfers of goods, including those used as inputs to manufacturing. Cascade taxes cumulate – inputs are taxed, and the outputs are taxed as well. For example, a tax on sheet steel used to make an automobile becomes part of the cost of manufacturing the automobile and the tax is itself again taxed when the automobile is sold. Cascade taxes were a precursor to the value added tax (VAT) systems. These have been replaced by VAT systems in most industrialised countries, but still apply in a few developing countries. Under a VAT system by contrast, each producer pays tax on only the increment to the value which has taken place since the last transfer, and it is the final consumer who pays the tax on the entire value of the good.

In the light of the foregoing, Brack *et al.* (2000) conclude that the definition in footnote 61, which on the face of it would appear to allow the BTAs for exports on energy/ carbon taxes, does not in fact apply to carbon or energy taxes. This conclusion, they maintain, is based on the following reasons:

1. The 1979 GATT Subsidies Code (which the ASCM replaced) referred to "goods that are physically incorporated in the exported product". The signatories of the Code had defined these as "such inputs [that] are used in the production process and are physically present in the product exported. The signatories note that an input need not be present in the final product in the same form in which it entered the production process".[12] It would appear to be most unlikely that the 1994 ASCM really intended to widen the scope for BTAs, a step which would be entirely contrary to the evolution of the multilateral trading system.

2. The possibility of a widening was noted during the Uruguay Round negotiations with some concern. A letter from an official of the United States Trade Representative's (USTR) Office indicated that the new language in the 1994 ASCM was the object of an informal agreement among developed countries whereby:

"… the change in question was proposed to address a specific and very narrow issue involving certain energy-intensive exports from a limited number of countries. It was never intended to fundamentally expand the right of countries to apply border adjustments for a broad range of taxes on energy, especially in the developed world… We discussed the matter with other developed countries involved in the Subsidies code negotiations. We are satisfied that they share our views on the purpose of the text as drafted and the importance of careful international examination before any broader policy conclusions should be drawn regarding border adjustments and energy taxes. The ongoing work on Trade and the Environment in the OECD may be particularly appropriate in pursuing this matter internationally."[13]

In the view of Brack *et al.* (2000), BTAs relating to production processes are only allowable if they are applied to inputs that are physically incorporated. BTAs appear not to be allowable if the input is not present in the final product – which is the case for energy consumed and carbon emitted during production.

It has been stated that the denial of the rebate for such taxes is consistent with the polluter-pays principles of ecologically sustainable development as it forces the producer to take responsibility for the environmental impacts of production. In this way, the GATT/WTO rules on the issue promote cost internalisation through environmental taxation. (McDonald, 2005).

Another group of commentators [see Pitschas (1995), Hoerner and Muller (1996), Biermann and Brohm (2003)] have stated (with differing reasons to support their positions) that under the 1994 ASCM, carbon/energy taxes on inputs that have not been physically incorporated into exported products are eligible for adjustment. Hoerner and Muller (1996) base their view on a construction of the negotiating history of the GATT.[14] Biermann and Brohm (2003), on the other hand state that there is no evidence of the existence of any written formal agreement (that would give the terms used in footnote 61 a meaning different from their ordinary meaning) as asserted by the USTR official, and that the legal effect of the letter would be of minimal value should a dispute involving the remission of energy taxes be referred for conciliation under the WTO system.

None of these views represents the final word on the matter. It would be up to a WTO dispute settlement panel to draw a conclusion on the issue.

5.5.3. Case studies of process-based BTA mechanisms in environmental excise legislation

This section examines features of process-based border adjustment mechanisms implemented in environmental excise taxes in the US – the Superfund Chemical Excises (Superfund Tax), and the Ozone-Depleting Chemicals (ODC) Tax.

5.5.3.1. The Superfund Tax

The *United States Superfund Amendments and Reauthorization Act of 1986,* (hereafter referred to as the Superfund Tax) was primarily designed to raise revenue for a trust fund devoted to the clean-up of contaminated toxic waste sites, where individuals who were responsible for the pollution could not be located. It was therefore designed as a rough attempt to place the burden of such clean-up on those responsible for generating the wastes, but was not intended to influence behaviour through the price system. The tax rate was modest: a few dollars per ton of the taxed chemicals and substances.

The Act created a system of taxes to fund the cleanup of toxic waste disposal sites, including a petroleum products excise, a corporate income tax surcharge, and a system of excise taxes on specified chemicals and derivatives.

The Superfund chemical excises applied to the sale or use of certain chemicals that were listed in a Schedule to the Act.[15] Its application also extended to untaxed chemicals manufactured using the taxed chemicals as feedstock. These chemicals together were described in the tax code as "taxable substances".[16] These derivates were not themselves subject to the excise taxes, but imports of them were subject to the tax, and exports were similarly rebated, as long as taxable chemicals constituted at least 50% of the chemicals used to produce the final substance, by weight or value.

As a result, imported chemicals were subject to BTAs equivalent to a tax on consumption collected at the level of the manufacturer: Imports were taxed on first sale or use by the importer, and any tax previously collected on exports was rebated. Because the tax was collected on the first domestic sale or use, a rebate of the tax on export was not always necessary.

The BTA mechanism in the legislation set up a three-tier system for determining the tax rate on an imported taxable substance. If the importer provided detailed information on the taxable chemicals actually used in the manufacture of the imported substance, the tax was based on the amount of tax that would have been paid if the substance had been manufactured in the US. Where the importer failed to provide sufficient information, a penalty rate of 5% of the appraised value of the product was imposed. The Secretary of the Treasury could, however, prescribe a rate – in lieu of the 5% – equal to an amount that would be imposed if the substance was produced using the "predominant method of production" employed in the US.

In the case of taxable substances that were exported, the manufacturer or importer was entitled to a refund in respect of the amount of the tax previously paid. As a precondition for receiving the refund, the party who paid the tax was obliged to either pay the refunded tax to the exporter or receive a waiver of such payment from the exporter. Alternatively, if the party who paid the tax waived the credit or refund, the exporter was allowed to receive it.[17] In either case, the refund or credit was based on the tax actually paid, and not on the tax imputed under the predominant method of production.

The significance of the BTA mechanism in the Superfund Tax was that the tax was calculated by reference to the amount of chemicals *used in the manufacturing process*; it was not necessary for all of the atoms contained in the taxable chemical to be physically incorporated into the final substance, and the tax rate was not adjusted if only a portion of the original chemicals were actually present at the end. Thus it was a true process-based BTA.

The tax was allowed to sunset at the end of 1995, owing to the inability of Congress and the Administration to arrive at a workable agreement on various aspects of financing the clean-up.

5.5.3.2. *The Superfund dispute and the GATT Panel decision*

The Superfund Tax was challenged by the European Community, Canada and Mexico before a GATT Panel.[18] At issue was whether these taxes that applied to imported products to offset the impact of domestic taxes on physically incorporated inputs to the like domestic products were consistent with the GATT? The European Community argued that the US tax was designed to raise money to clean-up pollution in the US caused by the chemicals during processing. As the imported chemicals had been processed abroad, the pollution in the country of production could be assumed to be taxed or subject to regulation in that country. Furthermore, chemicals exported from the US were exempted from the tax in spite of the fact that their production caused pollution in the US. The tax on imports was therefore inconsistent with the purpose of the Superfund Act and a principle requiring that polluters should pay. The Panel did not assess the consistency of the tax adjustments with this "polluter pays" principle, as it considered the principle as being irrelevant to a determination of the eligibility for BTAs under GATT law.

In endorsing the system of BTAs on taxable substances based on the actual consumption of taxable chemicals in their production, and the system of imputation based on the predominant method of manufacture, the Panel stated that the tax imposed by the US on imported substances was equal to the tax borne by similar domestic products as a result of the tax borne by their inputs. It was therefore consistent with the principle of national treatment.

The Panel found that a tax on "chemicals used as materials in the manufacture or production of the imported substances" was a tax "directly imposed on products" and therefore could be taken into account when imposing border adjustments on imported like products. It accepted the US argument that GATT contemplated the possibility for BTAs in respect of imported products that contained substances subject to an internal tax. The panel did not, however, indicate whether the chemicals needed to be physically incorporated in the final product in any way, or whether they could be consumed without remaining traces in the production process.

The Panel then referred to Article II:2(a) of GATT as clarifying that a tariff concession does not prevent the levying of "a charge equivalent to an internal tax imposed consistently with the provisions of paragraph 2 of Article II in respect of the like domestic products or in respect of an article from which the imported product has been manufactured in whole or in part". The Panel also recalled that the drafters of the GATT had explained: "If a charge is imposed on perfume because it contains alcohol, the charge to be imposed must take into consideration the value of the alcohol and not the value of the perfume, that is to say the value of the content and not the value of the whole."

It then ruled that the tax rates on imports under the Superfund tax were determined – in principle – in relation to the amount of the chemicals used and not in relation to the value of the imported substance. Given that the tax on imported substances was equivalent to the tax borne by the like domestic substances as a result of the tax on chemicals, the measure was consistent with the first limb of Article III:2.

The Panel expressed its concern about the fact that the enabling regulation under which the Secretary of the Treasury could prescribe a rate according to the PPM, had not been issued. It however concluded that given US tax authorities had discretion to dispense with the need for the imposition of the penalty rate – and in the light of express statements by the US that "in all probability the 5% penalty rate would never be applied" – the penalty rate provisions did not violate the first limb of Article III:2.

5.5.3.3. *The US Ozone Depleting Substances Tax*

To accomplish the phase-out of ozone-depleting substances under the terms of the 1987 Montreal Protocol, the US Congress enacted a system of excise taxes as well as production allowances for their manufacturers, as part of the *Omnibus Budget Reconciliation Act of 1989*. The tax was specifically designed to "permit market forces to aid the work of finding substitutes" for the taxed chemicals. It was intended to influence behaviour through the price system and was effective both in raising the price of taxed chemicals and in discouraging their production. The tax was phased-in gradually over a period of years.

The tax applies to a long list of ozone depleting chemicals at rates proportional to their ozone depleting potential, and increased year by year. The increases were ascertained by reference to a fixed amount. The predictability of the increases were probably as important to the incentive effect of the tax as the current tax itself, as it enabled both producers and consumers to plan ahead. A floor stocks tax was also introduced to ensure that stockpiles produced in one year would not be sold at a higher price in future years.

Exemptions were provided for recycled chemicals, exports, chemicals used as feed-stocks or in manufacturing rigid foam insulation, as well as other uses that did not pose environmental risks.

5.5.3.4. *The Border Adjustment Mechanism in the tax*

At the time that the proposals to enact an ozone depleting substances tax was under consideration, concerns were expressed by manufacturers and users about its potential to cause significant damage to the industry's international competitiveness. In addressing this, a BTA mechanism was integrated into the provisions of the legislation.

The BTA mechanism in the ODC tax has a number of features in common with those in the Superfund Tax. First, both BTAs are product and process-related. Second, imports and exports of ozone-depleting chemicals were taxed generally in accordance with the destination principle; as a result, the tax is imposed on imported products containing or manufactured with ODCs when the product was first sold or used by the importer, at rates equal to the domestic taxes; and exporters (or manufacturers) were eligible for rebates. This prevents foreign producers with no comparable tax from preying on US manufacturers, and enables US exports to compete on a level playing field with manufacturers located in other countries that have not introduced similar taxes.

For BTA purposes, the importer can compute the tax in any of three ways. As with the Superfund Tax, if the importer provides sufficient and accurate documentation of the precise composition of ozone depleting chemicals used in their manufacture, the tax imposed is based on the actual use. If the importer does not supply the actual production data, Treasury makes a determination on the basis of the amounts used in the predominant production methods of comparable goods in the US. Where comparable data is not available, Treasury is empowered to impose a 5% *ad valorem tax*.

The BTA also applies to all products which contain or are produced with ODCs, unless the Secretary of the Treasury determines that only a *de minimis* amount of such products were used in the manufacture or production. Treasury regulations have set the *de minimis* level at 0.1% of the cost to the importer of acquiring the product. The *de minimis* exception does not apply to refrigeration or air-conditioning equipment, aerosols or foams or electronic equipment.

5.5.4. *The US Ozone Depleting Substances Tax: an evaluation*

Unlike the Superfund Tax, the ODC excise tax was set at a relatively high rate; in 1994-95 for instance, the taxed price of CFC-11 and CFC-12 (the most commonly used CFCs) was roughly triple the untaxed price. Coupled with a high demand for CFCs as coolants in vehicle air conditioners, this fuelled the growth of substantial illegal imports of the substances into the US. In response to the problem, a range of strategies were put into action and this has succeeded in curbing the pace of the growth of the illegal trade, see Brack (1996) and (1997); Brack *et al.* (2000) and also Hoerner (1998).

Commenting on the US experience, Hoerner (1997) observed:

"The US experience with the ODC Tax, establishes the importance of BTAs to achieving the benefits of environmental taxation. As a result of the BTA system, the domestic ODC industry was protected from foreign predation while an orderly phase-out of ODCs was achieved. This enabled US chemical companies to play a leading role in the development of commercially viable substitutes for ODCs in a wide range of processes, substitutes of which are now in use around the world. The ODC tax was high enough that, without the BTAs, the domestic industry would have been rapidly extinguished by foreign imports, with no resulting benefit to the global environment. Given this market reality, the political reality is that the US Congress would never have enacted the ODC tax without BTAs."

A UNEP report indicates that by the end of 1998 (the latest date for which full data is available), production of the original CFCs had fallen by 95% in industrialised countries (the remaining production being devoted to essential use exemptions and exports to developing countries); and production of the original controlled halons had fallen by 99.8%. Although both production and consumption had increased in developing countries, as expected and allowed by the Montreal Protocol, overall world production had declined by about 88% (CFCs) and 84% (halons) from the base year, 1986. It notes that the growth in concentrations of the major ozone-depleting chemicals in the atmosphere has clearly slowed. The total combined abundance of ozone-depleters in the lower atmosphere peaked in 1994, and is now slowly declining.

5.5.5. *Lessons from the US's experience of BTAs under the Superfund and ODC taxes*

Hoerner (1998) draws the following lessons from the US's experience with BTAs:

- In the case of imports where the necessary information on the production process is limited or not provided by the exporter, the adoption of the predominant national method approach *in lieu*, would be compatible with the GATT rules.

- BTAs should be avoided where the tax is a trivial proportion of the price; the 50% floor for the Superfund Tax, and the *de minimis* rule for non-listed products for the ODC tax, helped to avoid the substantial administrative burden of calculating extremely small amounts of taxes with very little environmental benefit.

- If the tax reaches a significant proportion of the final price, there can be evasion, including the proliferation of illegal trade. This problem is not unique to BTAs, however; it stems from the relative prices and availabilities of the controlled products and their alternatives. The EU – without an ODC tax – experienced quite significant volumes of illegal imports. However, the use of taxes and BTAs are clearly likely to exacerbate such problems.

Goh (2004) cautions that there is a possibility that certain aspects of the *Superfund* decision that could have potentially negative impacts on competitiveness:

- The requirement that importers could only benefit from the normal tax rates if they provided the US authorities with sufficient information, raises concerns about the need to provide commercial and propriety information. He states that from a business perspective, the need to protect commercially sensitive and proprietary information – such as on the materials and processes involved in manufacturing a product – are very real considerations. The possibility that such information could fall into the hands of competitors, could have a very real effect in stifling trade.

- The mechanisms in the Superfund Act raised both trade and environmental policy concerns. As pointed out by the European Community, foreign competitors of the US producers of the taxable chemicals and substances could be assumed to have paid for the pollution caused by the production of chemicals and substances in accordance with a "polluter pays" principle. This would have been either directly (by paying a tax for the removal of pollution) or indirectly (by meeting regulatory requirements to prevent pollution). The measure therefore accorded US producers an unfair competitive advantage given that a chemical exported from the US was exempt from the tax under the Superfund Act and no corresponding tax was imposed when it was imported into the European Community.

Conversely, a substance containing the chemical exported to the US would have borne the costs of environmental protection twice – once in the exporting country in accordance with a "polluter pays" principle and upon import into the US under the US measure. The US measure in effect forced foreign producers to help defray the costs of US producers cleaning up the environment.

Goh (2004) notes that such measures are at variance with the 1992 Rio Declaration (principle 16) which obliges national authorities to internalize environmental costs and avoid trade distortions in the application of environmental measures.

5.6. The applicability of BTAs to carbon/energy taxes related to processes

Following the ratification of the Kyoto Protocol, a number of countries have introduced measures to address the challenges of climate change. In particular, the EU's emissions trading system came into force in January 2005, and a number of other countries have also introduced energy/carbon taxes. These developments have raised the question of whether carbon/energy taxes related to processes are eligible for border adjustment under the GATT rules.

5.6.1. Article XX (the General Exceptions Clause) of GATT

Article XX (the General Exceptions clause) permits limited and conditional departures from the principle of non-discrimination. As a result the violation of general GATT obligations (under Articles I, III, or XI) by certain measures would not necessarily be proscribed under GATT. PPM trade-restrictive measures might either be allowed under Article XX(b): where they protect human, animal or plant life or health; and are necessary for that protection; or under Article XX(g): where they are related to the conservation of exhaustible natural resources;[19] and are made effective in conjunction with restrictions on domestic production or consumption. Measures aimed at addressing Climate Change will fall within the scope of paras. (b) and (g) of Article XX. In WTO (1996d), a measure aimed at the conservation of clean air was found to fall within the ambit of para. (g).

In addition, the head note of Article XX imposes an overriding requirement that the measure must not be applied in a manner which would constitute a means of arbitrary or unjustifiable discrimination between countries where the same conditions prevail; or a disguised restriction on international trade.

In the *Shrimp-Turtle* decision, the Panel held that Article XX represents a balance between the rights of WTO members to implement measures to pursue objectives "recognised as important and legitimate in character" and the rights of other members to benefit from compliance with the GATT/WTO obligations.

Whilst both exceptions might arguably apply to carbon/energy taxes, or an emissions trading measure, it is quite conceivable that a dispute settlement panel could adopt a restrictive construction of the article. In such an event, it could construe the primary objective of the BTA as a discriminatory measure designed to protect domestic industry from the actual or perceived impact of the tax, rather than to mitigate the effects of climate change. This, however, is not a foregone conclusion until a dispute panel makes a formal determination of the issue. The intricacies of the underlying issues are demonstrated by the fact that the WTO's Trade and Environment Committee has been engaged in ongoing discussions of the issue since its establishment in 1995 and has not been able to resolve the issue one way or the other.

Notes

1. There is a divergence of views among governments and commentators as to whether energy taxes (as opposed to other environmental policies and measures) represent the most effective means of addressing climate change.

2. The countries are: Austria, Denmark, Finland, Germany, Italy, Netherlands, Norway, Sweden, and the United Kingdom.

3. See OECD (2001a) and Chapter 2 above for a more detailed description of current uses of environmentally related taxes.

4. To achieve significant reductions on climate change scenarios in the long term, the German Council on Sustainable Development for instance argues that Germany would need to reduce CO_2 emissions until 2050 by 80% compared to 1990 levels [See Biermann and Brohm (2003) and German Council for Sustainable Development (2001)]. In Australia, the Greenhouse Advisory Panel has recommended that the state of New South Wales slash its emissions by 60% by 2050, New South Wales Energy Green Paper, 2005.

5. Exceptions to this central principle are permitted in certain defined circumstances, such as regional trading arrangements or preferential arrangements for developing countries.

6. Much of the information in this section is based on OECD (1994).

7. See footnote 58 in Annex I to the Agreement on Subsidies and Countervailing Measures 1994.

8. For a detailed overview of the evolution of the concept, see OECD (1995).

9. The Panel Report was circulated in 1991, but was not adopted, as the US and Mexico arrived at an "out of court settlement". The process-product standard in this case therefore does not have the status of a legal interpretation of GATT law.

10. Fauchald (1998), at 242-244 makes the point that the timing of the assessment is also critical. A tax that applies equally to domestic and imported products at the time of introduction may only ultimately affect imports, since local producers whose primary market is local have a greater reason to adjust their practices to avoid the tax. If the protectionist application of the tax is assessed at the time of a complaint, it may well be shown to favour domestic producers, even though that was not the case when it was introduced.

11. See WTO (1996c) and WTO (1997a). These issues are discussed in detail in Fauchald (1998).

12. Guidelines on Physical Incorporation, Doc. SCM/68, paragraph 4, discussed in GATT (1986).

13. Letter from D. Phillips, Assistant United States Trade Representative for Industry, to A. Katz, President US Council for International Business, referred to in Inside US Trade, 28 January 1994. An excerpt of this letter is reproduced in OECD (1994).

14. The validity of this approach has been challenged as being inconsistent with the purported interpretation when construed in the light of the provisions of the Vienna Convention on the Law of Treaties, see Goh (2004).

15. The list of taxable imported substances (26 USCA 4661) expanded from the original list of 49 in 1987 to at least 123 by 1 January 1995. (Davie 1995). [Chapter 26, refers to the US Internal Revenue Code; the abbreviation USC stands for "United States Code"].

16. See 26 USC 4672(a).

17. See 26 USCA 4662(e) (2).

18. See GATT (1987).

19. "Relating to" has been interpreted as "primarily aimed at" conservation, without being necessary to it, see WTO (1996d).

ISBN 92-64-02552-9
The Political Economy of Environmentally Related Taxes
© OECD 2006

Chapter 6

The Sectoral Competitiveness Issue – *Ex post* Studies

W hile economists have set out the desirable features of market based instruments for environmental policy, those theoretical perceptions are rarely met in practice. Since governments cannot design and implement policy measures without taking account of political realities, there will usually be a disparity between theory and practice. First, while economists may concentrate on "optimal" instrument design serving one overriding goal – economic efficiency –, political reality may demand that other goals, which are not necessarily consistent with each other, also play a part in practical design and implementation. Second, governments are not all-knowing, all powerful guardians of social wellbeing. Rather they have to contend with pressure groups and lobbies which, in turn, represent sets of conflicts of interest. As a result, actual policy and "optimal" policy rarely coincide. This gap is very much subject of a political economy approach to policy analysis. One goal of political economy is to analyse this gap and the processes influencing it.

We will not give an in-depth political economy analysis here, but simply draw some lessons of political economy by analysing some empirical country case studies. We shall see that for instance the issue of international competitiveness, discussed in Chapter 4, often emerge in these political processes and may influence both the implementation and the design of environmentally related taxes. The same can be said of another issue raised in this publication; income distribution. However, the income distribution issue will be discussed in Chapter 7.

A main finding drawn from policy-making experience is that significant "competitiveness" pressures are indeed a reality in certain cases, depending on the type and design of a given environmentally related tax, and the characteristics of the markets and firms affected. While it is often said that it is difficult to find examples of environmentally related taxes having a serious negative impact on the competitiveness of any sector, it must be remembered that this situation results directly from provisions to protect industry (to date, primarily exemptions) that typically accompany the introduction of such taxes.

However, those strongly opposed to introducing environmentally related taxes on competitiveness grounds sometimes tend to forget that alternative policy instruments used to reduce pollution, such as regulations, also affect firm's costs and impact on the competitive position of individual sectors and the country as a whole. By enhancing the economic efficiency by which a given target is reached, properly designed taxes will help minimise adverse effects on competitiveness nation-wide – compared to *e.g.* direct regulation or "voluntary approaches".[1]

Furthermore, the opposition tends to overlook that environmentally related taxes are one of a number of factors determining a firm's overall competitiveness. Research on economic performance shows that skills and capital investment largely determine sectoral competitiveness.

First this chapter will present some empirical country case studies:

- the industrial energy consumption tax in France;

- the United Kingdom Climate Change Levy;
- the MINAS accounting system in the Netherlands;
- the Swiss heavy goods vehicle road use fee;
- the Irish plastic bag tax; and
- the Norwegian Aviation Fuel Tax.

We will then summarise the practical measures used to limit negative impacts on sectoral competitiveness, to end up with some political economy lessons that can be drawn from the case studies.

6.1. The industrial energy consumption tax in France[2]

A recent example of competitiveness obstacles met in attempts to realise environmental goals is that of France, and in particular, the attempt by the French government over the period 1999-2000 to tax energy consumption by industry (in order to help France achieve its Kyoto obligations). This example illustrates not only the force of industry concerns, but also the need to carefully select compensating measures that will not run foul of domestic laws or other legal constraints. As elaborated below, the final bill presented to parliament, which contained significant concessions to business, was ultimately *ruled offside* by the Constitutional Court.

Why the implementation of the energy consumption by industry tax?

At the national level, the energy tax project was part of the previous ecological tax reform initiated in 1999. The government decided on 20 May 1999, after consultation with the different administrations, the principle of a new energy tax which was to be implemented in 2001. This decision was presented as the consequence of the French commitment in Kyoto's agreement.

How was the energy consumption tax supposed to be implemented?

Under the original proposals, companies would have been subject to a tax on energy consumed as an intermediate input. Fossil energy sources were to be taxed according to their carbon content at the rate of EUR 40 per tonne of carbon equivalent. Households and certain sectors (agriculture, fishing, forestry, and the transport sector) would have been exempt. New firms would also have been exempt during their first year of activity.

During the development of the proposal, competitiveness concerns were raised by many sectors and focused mainly on two factors: energy intensity and openness to international markets. Sectors were characterized according to their openness to competition.[3] However, differentiating firms by sector is difficult, owing to "borderline cases" and the fact that certain firms have multiple production activities. Thus, the focus would be on energy intensity, with studies showing significant energy consumption per unit output (or per unit value added) in some sectors – notably the steel, fertilizers, minerals extraction, cement, and plastics industry sectors – and significant variation across sectors.

To alleviate competitiveness pressures from the energy tax at EUR 40 per tonne of carbon equivalent, the tax base would be scaled back by a factor – the tax-base reduction factor – varying with the energy intensity of the firm measured by energy consumed (in tonnes oil equivalent) per unit value added. The tax reduction factor increases with the energy intensity of the firm, as illustrated in Figure 6.1. For the most energy-intensive firms – those consuming 61 tonnes or more oil equivalent per million euro of value added – the tax-base would be reduced by 95%.[4] The tax-base reduction is noteworthy, as it runs counter to the basic operation of an energy tax that would impose a progressively higher tax burden per unit of output or value-added, as energy use increases.

In order to be granted a tax-base reduction, firms were obliged to negotiate "voluntary agreements" with government on emissions abatement. At the same time, by undertaking such agreements, firms were rewarded with *additional* tax reductions if their carbon emissions were lowered.[5]

Figure 6.1. **Tax base reduction coefficient of proposed carbon-energy tax**

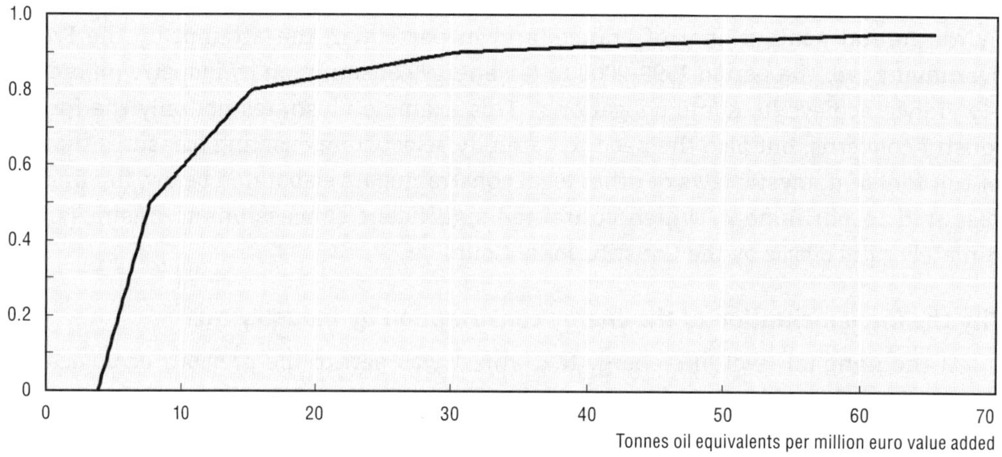

Tonnes oil equivalents per million euro value added

Source: Delache (2002).

Main problems/criticisms

The energy tax reform, voted by the Parliament in November 2002 was rejected by the Constitutional Court in December 2002. The court considered that the tax was against the "equal treatment" established in the French Constitution. On one hand, the tax base reduction run counter to the basic operation of an energy tax that would impose a progressively higher tax burden per unit of output or value-added, as energy use increases. Additionally, the design of reductions could allow more polluters to pay less tax.

Main lessons

Given these observations, an emerging lesson is that policy-makers should take steps to ensure that competitiveness pressures are adequately assessed and addressed. As the French example also illustrates, even where significant concessions are made, proposals may fail if the mitigating measures run foul of domestic (or possibly international) law.

Therefore, another observation is the importance of considering a short-list of mitigating options alongside legal obligations and possibly other constraints to ensure that they would not be found to provide *e.g.* a prohibited subsidy.

6.2. The United Kingdom climate change levy[6]

Why was the climate change levy implemented?

The climate change levy (CCL) and the UK carbon emissions trading scheme arose from the "Marshall report" in 1998. This report proposed two measures (an energy tax and a permit trading scheme) to act on climate change using market-based instruments in a difficult political context.[7] First, the new Labour government did not want to introduce measures that might have a disproportionate effect on the poor. Second, Labour owed an allegiance to the coalmining communities, in stark contrast to the previous government which had successfully made overt attempts to curtail the power of the mineworkers. Third, Labour had to escape a past image of "high tax and high public spending", so that whatever measures were introduced had to be as friendly to industry as possible and had to avoid the impression that any new tax was simply for raising revenue.[8]

How was the CCL implemented?

Taking on board these three main governmental concerns, the levy was designed to avoid taxing households, keep industrial cost burdens to a minimum, and bring industry on board with the UK climate change programme. The levy is "downstream", *i.e.* is paid by energy *users*, not *extractors* or *generators*, is levied on industry only (including agriculture and public sector), with households and transport being exempt (also renewable energy since 2002), and is structured so as to encourage renewable energy but not nuclear power (users of nuclear electricity pay the tax). An 80% discount could be secured if the industry in question negotiated a "Climate Change Agreement" (CCA) – *i.e.* an industry package of measures to reduce emissions relative to some baseline.[9]

Additionally, the CCL was designed as part of a revenue-neutral reform. Most revenues from the CCL are "recycled" back to CCL-paying industry in the form of reductions in employers' contributions to social security in order to encourage employment.[10] The introduction of these reductions was one of the main arguments used in the Marshall Report to ensure that the tax did not harm "competitiveness". Also, reinforcing the substitution effect arising from the resulting price increases, part of the revenues from the CCL are used to finance carbon technology improvements.

When first implemented, only the energy-intensive industries under the definition of the European Integrated Pollution Prevention and Control Directive could use CCAs to face lower rates of CCL. However, after a government's promise of extending the CCAs to other industries, in the 2004 Budget two additions were made. First, the qualifying "*energy-intensity threshold*" (ratio of energy expenditure to value of output) was reduced from 20% to 12%. Second, an "*international competitiveness test*" was also introduced: any sector with an energy intensity threshold in the range 3-12% will qualify for a CCA if, in addition, it has an import penetration ratio of 50% or more, or an export-to-output ratio of 30% or more.

Main results of the CCL

While the CCL was designed as part of a revenue-neutral reform, this does not mean that each and every industry would find itself in a tax-neutral position. ECOTEC (1999) estimated the effects of the early 1999 CCL proposed levy. This proposed levy was not equal to the eventual levy. The study suggested that the net effects of the CCL and the social security rebate would be very modest – either beneficial or a very small cost – for most industries. The exceptions were food products (net cost of GBP 87 million p.a.), iron and steel (GBP80 million); oil extraction (GBP 54 million), pulp and paper (GBP 48 million), and a few other industries. Even for these sectors, the net tax is very small relatively to sectoral turnover. Moreover, for those industries facing a net tax, most are energy-intensive and they could take advantage of the provision for a CCA.

Regarding the CCAs, ETSU (2003) found that over half of the CCAs had achieved their targets, although inspection of the data suggests that around a quarter of the agreements are adrift of their targets by 5% or more. It also notes that one effect of the CCAs has been to motivate corporations to raise energy saving up the corporate agenda, but the costs and benefits of the CCAs appear not to be known with any precision.[11]

Agnolucci et al. (2004) provide an assessment of the emissions effect. They define the general effect of the tax as comprising an announcement effect and an implementation effect (see also Agnolucci and Ekins, 2004). The announcement effect tries to account for industrial responses to the fact that the tax has been announced and will be implemented. The implementation effect accounts for the residual, i.e. any effect not accounted for by the announcement effect.[12] Using regression analysis, Agnolucci et al. (2004) test for the presence of an announcement effect in three broad sectors: the industrial sector, the commerce and other final users sector (basically commerce and public administration), and the economy as a whole. They find an announcement effect (i.e. energy reductions in 1999-2001, the period between the announcement of the tax and its introduction) only for commerce and other final users. Agnolucci et al. (2004) conclude that:

> "This is a positive result as it shows that a credible Government policy of pre-announcing new taxes can lead to early action by firms. The study has shown that this change is ... permanent and not transitory".

However, the effect is found only for the commercial and other users' sectors which, together, account for only 10% of final energy demand or 15% of non-household final energy demand. The announcement effect appears to be absent for industrial and economy-wide demand. This does not mean that the CCL package has not had a general price effect but, unfortunately, Agnolucci et al. do not estimate what this might have been.

On the other hand, an assessment of the CCA component of the CCL package by UK DEFRA regards the achievements of the CCAs up to the end of the first target period ending 2002 as "massive".[13] The long run goal of the CCAs was to achieve a 2.2 MtC reduction in emissions by 2010. Yet the 2002 figures suggest a reduction of 3.7 MtC from the baseline, implying substantial over-achievement well before the final target date.[14] As mentioned in Chapter 3, similar results were found in a recent analysis made by Cambridge Econometrics (2005). They estimated that total CO_2 emissions were reduced by 3.1 MtC – or 2.0% – in 2002 and by 3.6 MtC in 2003. The reduction is estimated to grow to 3.7 MtC – or 2.3% – in 2010.

One problem when implementing the CCA was the "additionality" of the CCA achievements, i.e. whether they represent energy/emission reductions that would not otherwise have occurred. The original attempt in the CCAs to prevent non-additionality involved estimating "Business-as-Usual" energy consumption figures, then setting a limit target that would have involved all conceivable cost-effective measures being undertaken, and finally setting the actual targets to be 60% of the gap between business as usual and these limit targets. The 2002 targets were then calculated as a step on the way to this "60% gap closure".

Main problems/criticisms

The main criticism of the CCL is its disparity with respect to a pure carbon tax. Table 6.1 translates the CCL rates of charge into an implicit carbon tax by taking account of the carbon content of fuels. Since the tax per kWh of gas and coal is the same, and given the lower carbon content of natural gas, the effect is to produce a differential tax rather than a uniform one. Hence the tax is potentially inefficient as long as there are opportunities to switch between fuels. The carbon content of natural gas is only 54% of that of coal.[15] While one would therefore expect encouragement of a switch to lower carbon use fossil fuels, the CCL does not encourage gas to be supplied at the expense of coal. As originally formulated, the flat rate for electricity, however it was generated, provided no incentive at all for electricity users to switch to cleaner fuels, and no incentives to use energy more efficiently (since the tax is charged per unit of electricity). This problem was ameliorated slightly in the November 1999 revisions, since renewable energy and CHP were then exempted.

Table 6.1. **Translation of climate change levy tax rates to a pure carbon tax**

	Original rates	Nov. 99 rates
Coal	GBP 23 per tonne carbon	GBP 16 per tonne carbon
Natural gas	GBP 42 per tonne carbon	GBP 30 per tonne carbon
Petroleum	GBP 31 per tonne carbon	GBP 22 per tonne carbon
Electricity[1]	GBP 43 per tonne carbon	GBP 31 per tonne carbon

1. Based on the prevailing fuel mix for electricity generation.
Source: ECOTEC (1999).

Administrative simplicity is one argument for not reflecting carbon intensities in the CCL. Since the tax is levied "downstream" it is not possible, the argument goes, to identify the fuel sources "belonging" to any given unit of electricity. This is largely true, but only arises because of the way the tax was designed. There was, in principle, no difficulty in designing a pure carbon tax. At the time, the CCL was also argued to provide "fair competition between fuels" [see HM Customs and Excise (1999), para. 5.3], but this appears to involve a very narrow definition of fairness, since if one fuel pollutes more than the other, economic efficiency would demand that it be taxed at a higher rate. Again, the justification really reflects the concerns at the time about the impact on the coal industry.

Another important point relates to the exclusion of the transport and household sectors. The exclusion of transport can be explained by the fact that, at that time, separate market-based instruments were being directed at the transport sector. The fuel duty escalator (FDE) committing to a yearly increase in the fuel duty by 3% per annum was in place, although its foundations by this time were shaky.[16] Other differential fuel taxes were coming into being, along with company car tax reform, changes to vehicle excise duty, and so on. But the recent experiences of great opposition against the FDE indicated that imposing a further tax on transport would risk political protest on a significant scale. Excluding transport was therefore as much a political necessity as anything else. Arguably, environmentally related taxation of road transport was also adequate in terms of the marginal damage costs of carbon were concerned. As noted earlier, the exclusion of the household sector also had a political motivation, but additionally reflected the goals of government as set out in a 1997 Treasury statement that environmentally related taxation must be "well designed, meet objectives without undesirable side-effects, keep compliance costs to a minimum, have acceptable distributional impacts, and take account of implications for international competitiveness" (see HM Treasury, 1997). The exclusion of the household sector from the CCL would therefore have a justification because of the probable regressive impacts of an energy tax.[17]

Nuclear power plays an important role in the CCL. Nuclear power is not exempt under the CCL despite being a zero carbon energy source.[18] This is because, despite its name, the CCL is an energy tax, not a carbon tax. Had it been a carbon tax, the competitive position of nuclear power, already seriously compromised in the UK because of price falls in the liberalised energy markets, would have been enhanced. But had nuclear been exempt, the CCL would have been seriously compromised in the sense that revenues would have been drastically affected, and coal would have been exposed to further contraction. Given that revenues from the CCL are recycled (used to reduce employers' social security contributions) and hypothecated (used to finance carbon technology improvements), the revenue raising effect of including nuclear power would appear to be irrelevant. But including nuclear power nonetheless increased revenues, enlarging the fund available to "buy in" cooperation from industry. One other argument for including nuclear power in the CCL has been advanced in some quarters. Had nuclear power been exempt, it would have secured a competitive advantage that would have favoured imports of electricity from France via the interconnector. In short, while including the sector in the CCL was clearly going to damage UK nuclear power, excluding it would only be to the advantage of France. Nowadays, British Energy continues to exert pressure on government, arguing that it could save GBP 80-100 million p.a. if it was exempt from the CCL.[19]

The "revenue effect" expected from the use of part of the tax revenues to encourage longer-run switches to low carbon technologies has also been questioned. The very modest sums (around 4% of CCL revenues per annum, GBP 24 million in 2001-02 and GBP 36 million in 2002-03) spent by the Carbon Trust[20] suggest that the use of such revenues to secure long-run substitution effects is unlikely to be effective. However, the Trust cannot be seen in isolation since the CCAs themselves also have the effect of shifting energy conservation and hence, to some extent, lower carbon technologies to the fore.

However, the real problem of the UK climate change policy is its incompatibility with the EU ETS. Anyone over-complying with the CCA agreement could, in principle, trade the resulting credits into the UK emissions trading scheme (ETS), along with permits allocated under that scheme and renewable energy certificates under a separate renewable energy constraint on generators.The linkages between the UK CCL/CCA package, the UK ETS and the EU ETS are immensely complex and they are out of the scope of this publication, for an excellent analysis see Sorrell (2002).

Main lessons

This case study suggests that in practice environmentally related taxes diverge substantially from the textbook ones.

Regarding competitiveness concerns, the case study supports the views that the general rule in implementation of such as taxes is to set provisions to protect industry (primarily exemptions). It also suggests, as stated in the above section on theoretical case studies, that the assessment of the effectiveness of the environmental-related taxes and their impact on competitiveness widely depend on the benchmark used to judge the effectiveness of the tax.

Finally, another important lesson is that policy-makers must ensure that the design of environmentally related taxes must be compatible with the environmental international legal framework. Otherwise the credibility of environmentally related taxes can be threatened with the danger of being catalogued as another revenue raising instrument.

6.3. The MINAS accounting system in the Netherlands[21]

Why was the MINAS implemented?

The nitrogenous (N) and phosphorous (P) accounting system MINAS was the core instrument of the third phase of the manure policy in the Netherlands,[22] and marked a shift in the manure policy from ineffective command-and-control regulations and measure-oriented policies (implemented in the 1980s and first half of the 1990s) towards target-oriented policies with economic incentives.

MINAS was a farm-gate balance approach for nutrient accounting that regulates nitrogen (N) and phosphorus (P), which was implemented in the Netherlands in 1998.[23] The basic reason for implementing this accounting approach was the need to regulate N and P from both fertilizers and animal manure (as Figure 6.2 shows, there is strong correlation between livestock density, N fertilizer input and N surplus per ha; countries with the highest livestock density also have the highest N fertilizer input and highest N surplus on the soil surface balance). Further, there was the need for an instrument that provides an incentive for good nutrient management at farm level. MINAS was seen as both a management instrument for farmers to improve nutrient management at farm level and as a regulatory instrument for the government to regulate N and P losses from agriculture to the wider environment, *i.e.* to groundwater, surface waters and atmosphere. It was also seen as a

Figure 6.2. **Total net nitrogen input on agricultural land
and mean nitrogen surplus**

Includes inputs via fertilisers and animal manure, 1997

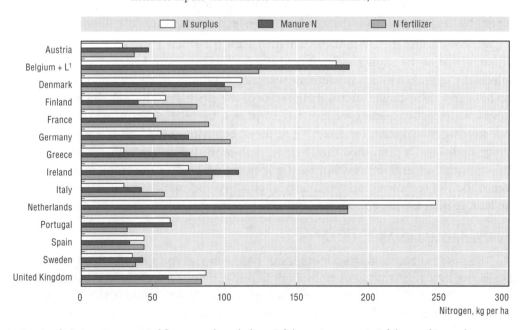

1. Inputs of nitrogen are corrected for ammonia emissions. Belgium + L represents Belgium and Luxembourg.
Source: OECD (2001b), CEC (2002c) and Eurostat (2002).

flexible instrument that would be able to address the large differences between farms in environmental performance.

How was the MINAS implemented?

MINAS involved registration of all N and P inputs and output at farm level, using a farm-gate balance sheet approach. Inputs of N and P via fertilizers, animal feed, animal manure, compost and other sources, as well as the N and P output (export at farm level) in harvested products, including any animal manure have to be recorded accurately, using official documents for sales and purchases from accredited firms only.

The difference between total N and P inputs and outputs should not exceed certain "acceptable" (levy-free) surpluses for N and P. When N and/or P inputs exceed the N and/or P outputs plus the levy-free surpluses for N and P, farmers pay a levy which is proportional to the excess above the levy-free N and/or P surpluses. Levies provided the incentive to both lower the import of N and P in fertilizers, animal feed and/or animal manure, and to increase the export from the farm of harvested products and animal manure. Levy-free surpluses were differentiated according to soil type and land-use, and were lowered step-wise between 1998 and 2003. The levy-free surpluses were a compromise between agronomic and environmental objectives and interests, as they were the result of political debate in the Parliament.

In principle, farmers had the choice of decreasing the surpluses to the level of the levy-free-surpluses and pay no levy, or to decrease the surpluses less and pay a levy in proportion to the exceedance of the levy-free surpluses. In practice however, the levies were meant to be prohibitive, and taking measures was more economical than paying levies.

In 1998, levies were EUR 0.68 per kg per ha for N and EUR 2.6-10.4 per kg per ha for P. In 2002, levies were raised to EUR 2.53-5.07 per kg per ha for N and to EUR 20.60 per kg per ha for P, which is about 5 to 10 times the price of fertilizer N and 50 times the price of fertilizer P, respectively. Levies were increased in 2002 to make them prohibitive, in response to queries by the European Committee about the effectiveness of the initial levies.

MINAS was designed with the aim of increasing the efficiency of nutrient utilization at all farms, and at the same time to drastically decrease nutrient losses. This basic design reflects the approach of eco-modernists who suggest that over-consumption of resources can be overcome by drastic increases in resource use efficiency. Through the application of a levy to surpluses exceeding the levy-free surpluses, the MINAS instrument was perceived as a steering mechanism for directing farmers' behaviour along more environmentally benign pathways. It presumed that farmers' behaviour is economically rational and that the economic incentive provided by the levy would lead to the implementation and acceleration of technical and technological changes and management improvements at the farm to increase the efficiency of nutrient utilization.

Main results of MINAS

The nutrient accounting system MINAS has been shown to be an effective policy for decreasing total nutrient losses from dairy farms. Potentially, it could also be an effective instrument for arable farms, but there is no clear evidence yet. However, MINAS turned out to be not effective and efficient for pig and poultry farmers.

It has been estimated (RIVM, 2002 and 2004) that 80% of the decreases in N and P surpluses during the period 1998-2003 was the result of the manure policy and 20% of other polices and autonomous developments, which made MINAS in practice a quite effective instrument for decreasing nutrient surpluses from dairy farming.[24]

However, MINAS was considered not to be a proper instrument for pig and poultry production (RIVM, 2002; 2004). It was fundamentally flawed for intensive animal production systems that have little land and import basically all animal feed, because MINAS expressed surpluses in units per surface area, and did not address variations in nutrient stocks in animals, animal feed and animal manure. A lot of effort was made in practice to improve the accuracies of sampling and analyses, but the inaccuracies that remained were larger than MINAS could afford.

Furthermore, MINAS did not prove to be effective for arable and horticultural farming, simply because P fertilizer was excluded from accounting in this sector, the fixed defaults for N and P in harvested crops were set too high, and N and P surpluses at most farms were already below the levy-free surpluses. In theory, MINAS could also be an effective and efficient instrument for arable farming, if the fixed default values for N and P in harvested crops were made crop-rotation specific, P fertilizer was included in the accounting and the levy-free surpluses were adjusted to environmentally sound levels.

Additionally, the economic cost of enforcement and administration of MINAS and other instruments have been increasing greatly over the last few years. The high cost of administration was in part caused by fraud, exploitation of loopholes, legal procedures, and by the many changes that were made in the MINAS system and in additional policy instruments. Through these many changes, there was insufficient time for proper implementation and fine-tuning of MINAS in practice. As such, MINAS did not receive the credits that it deserved as instrument to decrease nutrient losses from especially dairy farming, according to the polluter-pays principle.

Though the manure policy greatly affects agriculture in the Netherlands, it would be naïve to ascribe all changes in agriculture to the manure policy and all effects of the manure policy to MINAS. Agriculture in the Netherlands is heavily influenced by changes in the (export) markets for agricultural products in general and by (the reforms of) the common agricultural policy (CAP) of the EU in particular. This complicates assessment of the effects of the manure policy in general and those of MINAS in particular.

Main problems/criticisms

The MINAS presumption about the relationship between the size of the levy-free surpluses and the size of N and P losses to the wider environment was often criticised (*e.g.*, Schröder *et al.*, 2003). In the long-term, in a steady-state situation, N and P surpluses according to MINAS are indeed a proper indicator for N and P losses to the environment. In the short-term this is not the case, though there are differences between N and P. Farmers often find it frustrating that their efforts to decrease N and P surpluses have so little direct impact on environmental quality. The lack of a direct relationship between N and P surpluses in agriculture and N and P losses from agriculture was also basis for scepticism about the environmental effectiveness of MINAS and at the same time the main argument in the political debate for decreasing levy-free surpluses less than initially planned.

There were a large number (about 40) of alterations and changes in MINAS since its implementation in 1998, and not all of these changes made MINAS more effective and transparent; in some sense they also led to confusion by farmers.[25] These changes were made in response to increased insights over time, complaints by farmers and political lobbying. For example, these changes include changes in the size of the levy-free surpluses and the size of the levies, but also in changes of default values and various correction factors (*e.g.*, RIVM, 2004; Schröder *et al.*, 2003).

Though MINAS was an integrated and flexible instrument and in principle applicable to all farms, the differences between sectors are so large and the manure problem is so complex that there is no single instrument that can solve everything at once. Additional instruments have been implemented to support the manure policy, but not all of these additional instruments were effective and supportive. Furthermore, the many additions and changes in the manure policy often led to confusion and disbelief among farmers (and policy makers), which further complicated its effectiveness.

An European Court judgement in 2003 declared MINAS against the EU laws mainly because of not being a command-control mechanism.[26] This judgement together with the increasing administrative burden, increasing fraud, and the absence of environmental benefits in the pig and poultry sectors (where manure surpluses are largest) led to a very rapid erosion of MINAS. Therefore, by the end of 2005, MINAS was replaced by a (complex) system of application limits for animal manure and fertilizers, in compliance with the Nitrate Directive. It remains to be seen whether this new system is equally effective and efficient for dairy farms, arable farms and pig and poultry farms.

Main lessons

A lesson to draw from this case study is that the design of the environmental economic instrument is important, particularly in the long-run, for the instrument to be effective and give incentives for behaviour change. In theory, a nutrient balance at farm level is the best instrument for decreasing total losses of nutrients from agriculture. It is meant to be fair, because it directly addresses the nutrient surpluses (polluter-pays principle). It is meant to be environmentally effective, because resource use is improved and thereby total nutrient loss decreased; there are no substitutes for nutrients so as to circumvent possible levies. It is meant to be economically efficient, as it is up to the farmer to decide which measures are cost-effective, practical and convenient to lower the nutrient surpluses to the level of the predetermined levy-free surpluses. However, it does not provide an economic reason for farmers to continue to implement changes and innovations aimed at further decreasing surpluses once levy-free surpluses have been achieved.

Another lesson is that providing information and awareness of the environmental problem and its alternative solutions before, during and after the implementation of the environmental policy is important for the acceptance and the effectiveness of the policy. The results of the fact-finding studies (RIVM, 2002; 2004) indicate that farmers need time to adopt new techniques and management styles to adjust to improved farming practices following the implementation of MINAS. They have to learn and they have to be convinced of the need for change, otherwise they remain reluctant to change and ignorant of improved practices. Direct guidance, demonstration farms, pilot farms are essential in this respect. On the other hand, government and governmental institutions also need time to learn and to fine-tune the instrument to the reality and complexity of the real-world (institution building).[27] Evidently, there need to be a balance between the step-wise lowering of the levy-free surpluses and the possibilities of farmers to adjust farming practices so as to lower the nutrient surpluses to the level of the adjusted levy-free surpluses.

Whilst the use of nutrient balances in agricultural research has a history of at least one century, using nutrient balances with levies on surpluses as an instrument to ecologically transform agriculture had no precedent. MINAS gave farmers the responsibility and the freedom to choose the optimal way to realize the targets of the manure policy and to solve the nutrient problem of their farms. However, initially, the nature and complexity of the manure problem was not understood well, there was certainly no consensus about the severity of the problem and there was also no consensus about the instruments needed to solve the problem. Moreover, there was and is still a lack of understanding of the relationships between "type of policy instrument – behaviour of farmers – agronomical and environmental effects". The response of farmers to the implementation of manure policy and measures appeared to be more varied and complex than initially expected. In retrospect, there has been quite some trial and error in the manure police of Netherlands.

6.4. The Swiss heavy goods vehicle road use fee

Why was the Swiss heavy goods vehicle fee implemented?

The history of the distance related road-use fee for heavy goods vehicles started in the 1970s and it was implemented in 2001. In between lays an eventful history with setbacks and several referenda.

The underlying idea was that heavy goods transport should pay for all the costs it generate. It was also part of a policy to move more transport from road to rail. A first step for a heavy goods vehicle fee was made in 1985 with the introduction of a flat fee for heavy goods vehicles. The flat fee was an intermediate solution to be replaced by a distance-related fee in the years to come. An important step in this direction was made in 1994, when a large majority of the Swiss people accepted the constitutional basis for a distance-related fee in a referendum. Two years later, however, the realisation of the project seemed beyond hope: the outcome of the consultation about the proposal for a law that was necessary to implement a distance – related fee was negative and tactical manoeuvres against the envisaged change had stopped the technical work essential for a successful implementation. A change in the political environment, utilized successfully by the promoters of the fee, brought a change in the trend: on 27 September 1998 the proposal for the law, redrafted according to the critique in the consultation was accepted by 57% of the Swiss people. The fee was then introduced 1 January 2001.

How was the Swiss heavy goods vehicle fee implemented?

For competitiveness reasons the fee had to be levied on all heavy goods vehicles using Swiss roads, both foreign and domestic. Thus, an important step towards gaining acceptance for the fee was making a bilateral agreement with the EU, the EU being the most important user of Swiss road system. One of the conditions of the EU to agree to such a treaty was the raising of the weight limit for heavy goods vehicles from 28 to 40 tonnes. Due to fear that Switzerland might be invaded by an avalanche of 40-tonne trucks when raising the weight limit, Switzerland had so far refused to make such a change. However, the change from a flat fee with a low rate to a distance-related fee with a much higher rate was considered a suitable means to make up for the raising of the weight limit and therefore meet the needs of Switzerland as well as of the EU.

The fee was introduced in 2001 at the same time the weight limit was raised from 28 to 34 tonnes. The rate was differentiated according to the emissions and the vehicles were put into three categories according to their emission standards. The following rates were applicable for the years 2001 to 2004:

- Fee category 1(corresponds to emission class Euro 0): 1.15 ct/tkm.

- Fee category 2 (corresponds to emission class Euro 1): 1.0 ct/tkm.

- Fee category 3 (corresponds to emission class Euro 2 and better): 0.85 ct/tkm.

As these rates refer to tonne-kilometres, the fee can be calculated by multiplying the rate by the distance travelled in Switzerland and the maximum permitted laden weight of vehicle and trailer.

It was agreed with the EU to introduce the new fee in several steps and in parallel with the increase in the weight limit. Both the rate and the weight limit have thus been gradually raised. In 2005 the weight limit was 40 tonnes.

Main results of the heavy goods vehicle fee

Introduction of the fee has had an impact on road traffic and it has contributed to a stop in the former growth trend in number of vehicle km of heavy goods vehicles. Also the transit traffic has been reduced and the number of lorries crossing the Swiss alps is about 8% lower than in 2000 before the existing fee was introduced. Due to the reduction of the number of vehicle-km and the renovation of the fleet external costs were reduced. Calculation by the Swiss Agency for Environment indicates that in 2007 emissions of pollutants (CO_2, NO_2 and PM 10) from heavy goods vehicles will be about 6-8% lower than they would have been if the old regime had been maintained.

One of the reasons behind implementation of the fee was a wanted change in goods transport from road to rail. The expected change has not occurred so far. The competitive advantage gained by rail due to the introduction of the fee was more or less balanced out by a productivity gain of the road sector as a result of the higher weight limit.

Main problems/criticisms

The arguments and the opposition to the proposal were mixed and dispersed and were mainly expressed in connection with the referenda. Among the arguments put forward during the long process of implementation were:

- The fee should not be based on distance but on fuel consumption and also on emissions.

- The rate was too high and would impose too large costs.

- The fee would neither lead to a transfer from road to rail nor would it work as a catalyst for the integration of Switzerland into the EU.

- The calculations were not solid enough.

- The cantons demanded a larger share of the income of the fee.

- The system of collecting the fee (the On-board Unit) was not yet available on the market. Also the Swiss Government could not declare the installation of an on board unit mandatory to foreign vehicles.

The Government to a large extent answered the criticism by, for instance making their own calculations of additional cost that were considerably lower than other calculations made by opponents. It also managed to find practical solutions to calculate the fee also for foreign vehicles.

Main lessons

From the Swiss case, OECD (2005a)[28] points at the following experiences that can be considered of general value:

- It is important that the project is a part of a policy that is acknowledged as such in public. This policy of aiming at transferring goods from road to rail has been accepted by the Swiss people in several referenda.

- The acceptance of a project can be increased if the revenue is distributed back to the transport sector. This should of course be in line with the developed policy.

- The authority in charge has to offer a technical solution that is simple, reliable and suited to solving the problems resulting from the political guidelines.

- The calculation of the fee is based on external costs of road transport of heavy goods vehicles. According to opinion polls this was well accepted by the Swiss population.

In addition, an element that may have contributed to its success was that the fee was introduced gradually from 2001 and both the rate of fee and the weight limit was increased gradually. Gradual phasing in of taxes can soften the immediate cost impact and give companies time for production adjustments to avoid the tax.

Another lesson that can be learned from the Swiss case is the importance of seizing the right moment for pushing through a delicate project on the political agenda, *i.e.* when the circumstances are favourable. But it also shows that it is essential to have done the necessary basic work when the opportunity arises. The positive experiences with the Swiss heavy vehicle fee show that road pricing is a suitable means of managing transport demand.

The Swiss example also shows how to overcome challenges of different sorts that are faced when levying a fee both on domestic and foreign vehicles. It was crucial to establish a system that made it possible and practicable to levy the fee both on domestic and foreign vehicles to avoid unfavourable competitive consequences on domestic vehicles. The challenges in this respect both included overcoming some technical challenges of facilitating two different systems to calculate the mileage and making an agreement with the EU regarding the fee (including the weight limit). The technical challenge was a consequence of Switzerland not being able to declare the installation of an *on board unit* measuring mileage mandatory for foreign vehicles. There had to be an additional system in place to declare the fee based on mileage for foreign vehicles. The bilateral agreement with the EU shows that it sometimes can be useful or even necessary to make cross border agreements to be able to implement market based measures addressing environmental problems. The need for international cooperation is, however, more obvious when taxes are imposed on products or key factors of production where goods are traded widely in an international market.

6.5. The Irish plastic bag tax

Why was the plastic bag tax implemented?

Litter constitutes a significant environmental problem in Ireland, and plastic bags used to constitute some 5% of all litter by weight, according to Litter Monitoring Body (2004). The plastic bags' share of the eyesore was further increased by their long durability, the frequent strong winds in Ireland and a landscape dominated by hedgerows – where plastic bags, according to Convery *et al.* (2005), often would be "caught up" – and highly visible. Because of their essentially inert nature polymers do not biodegrade in landfill. While such longevity may be viewed in itself as a negative feature, polymers, including plastic shopping bags, do not contribute much to the more acute sources of environmental impact associated with landfill disposal – discussed further in Section 2.3.4.

In January 1999 the Irish Government produced a study on plastic shopping bags which examined the effect on the environment from use of plastic bags which were being supplied free of charge to customers by retailers.

The use of disposable plastic shopping bags had been a feature of Irish retailing for at least the past quarter century or so. Two related matters had arisen:

- the issue of litter; and
- a growing public awareness of, and concern for, the environment.

The estimated total number of bags consumed annually in Ireland was 1.26 billion, equivalent to an average consumption of one bag per inhabitant per day.

How was the plastic bag tax implemented?

A study was undertaken and several policy options to address the environmental problems created by plastic bags were considered. Having assessed a range of policy instruments it was considered that a levy of some form offered the most appropriate means of reducing consumption of plastic shopping bags and thereby reducing consequent environmental problems. The preferred option in that study was for a supply-based levy charging in excess of EUR 0.05 per unit on plastic shopping bags destined for use in the Irish market from whatever source.

In March 1999 the Minister for the Environment and Local Government produced an aide-memoire for the Government on the subject of plastic bags and their effects on the environment. The Minister sought Government agreement to the publication of the study for the purposes of public consultation and agreement that options for a levy system should be assessed. The Government noted the contents of the study and the Minister's intention to publish the study and also agreed that the options for a levy system should be assessed with a view to the design of an appropriate supply-side measure to reduce bag consumption and litter.

In June 2000, the Government agreed to the introduction of a supply-based levy at the rate of EUR 0.15 on plastic shopping bags manufactured in Ireland or imported for use in the Irish market to be administered by the Revenue Commissioners (the Irish tax authority). It was also agreed that revenues generated were to be hypothecated to a central fund for use in a manner to be determined by the Minister for the Environment and Local Government. Following on from the Government's decision an extensive consultation process commenced.

A supply-based levy was considered to be the simplest and most cost-efficient method of collection. The numbers involved would be limited and therefore more manageable. However, having considered the representations from all interested parties, the Government revised its position from a supply-based levy to a point-of-sale levy, with the retailer being responsible for the charging of the levy to the customer and the payment of the levy to the Revenue Commissioners.

The detailed operational design and legislative structure of the levy is as follows:

- The imposition of the levy to consumers at the point of sale of goods or products to be placed in bags, or otherwise of plastic bags in or at any shop, supermarket, service station or other sales outlet.

- The tax rate was set to EUR 0.15 per bag. The tax rate was *not* set based on an estimate of the value of the negative environmental impacts caused by the bags, but instead at a level believed to be sufficiently high to trigger a major behavioural change.

- Bags used for the protection of certain foods are exempted, so are reusable bags that are sold for EUR 0.70 or more.

- The retailers are required to pass on the charge to the customer, and to itemise the levy on the invoices, etc., issued.[29]

- The Revenue Commissioners is the collection authority, and they are powered to estimate and collect amounts due in the absence of returns or where the returns submitted does not reflect the true liability of the retailer.

The Revenue Commissioners used its information systems to assist the Department of the Environment and Local Government in implementing the levy. The Commissioners identified all persons who would be likely to supply plastic shopping bags based on the NACE sector codes recorded on their VAT system. Information leaflets on the levy were sent to all VAT registered traders. This was accompanied by an extensive multi-media advertising and public information campaign. The levy was introduced with effect from 4 March 2002.

Main results of the plastic bag tax

The Irish plastic bag tax has received widespread public endorsement with little in the way of opposition or contravention. It has had a dramatic impact on consumption of plastic bags and on the problem of visual litter. This initiative won the widespread support of the public and caught the imagination of people in many countries around the world. Quite apart from the immediate objective of cutting down on consumption of disposable plastic bags, it has been very effective in raising awareness of waste management issues and the part every citizen can play in reducing the amount of waste we produce.

The Plastic Bag Levy raised EUR 13 million in revenue in 2003. These proceeds go to the Environment Fund to support waste management and other environmental initiatives. Most retailers report a reduction of over 90% in the consumption of disposable plastic bags since the levy's introduction.

The levy has triggered a reduction in the use of plastic bags by more than 90%. According to Litter Monitoring Body (2004), plastic bags now constitute approximately 0.25% of litter pollution. Convery et al. (2005) refers to litter surveys conducted by Irish Business Against Litter and the National Trust of Ireland, indicating that the number of areas in which there is no evidence of plastic bag litter increased 21% between January 2002 and April 2003, while the number of places without traces of plastic bags had increased by 56%. Given the durability of plastic bags, these numbers seem remarkable.

A recent survey suggests that in excess of 90% of the public support the levy. Public opinion seems to credit the tax with having substantially reduced the litter problem within a short time span. Among the many reasons given in support of this is that the levy is better for the environment, it has eliminated the problem of plastic bags litter on the streets and also that re-usable bags are actually more convenient and hold more shopping. The fact that the revenues (over and above the administrative costs) are earmarked to a separate Environmental Fund might also have contributed the public's good perception of the tax – although there are also arguments against the use of earmarking. According to Convery *et al.* (2005), the levy has proven so popular with the Irish public that it would be politically damaging to remove it.

Main problems/criticism

Irish-based importers and distributors made a strong case for their position that a supply-based levy would not be workable. Their main objection to their proposed role as collectors of the levy was the level of the charge they estimated to be a tax at an average rate of 1 500% per bag. They also suggested that retailers would source their bags in other EU states and that smuggling of bags would become an issue.

Retailers broadly supported the introduction of the levy. However, their principal concern was the issue of the type of bag that should be charged and or exempted from the charge. In addition they expressed fears that the customer would refuse to pay the levy and that the retailer would be liable for payment of the levy.

Main lessons

As in the case of the Swiss heavy goods vehicle fee, this case study shows that it is important that the policy is widely acknowledged among the public. This was indeed the case for the Irish plastic bag tax.

The case study of the Irish plastic bag tax shows the importance of doing thorough initial research and to carefully consider other relevant policy options. Introducing a tax is not always the right answer. This study assessed several policy options/instrument to address the environmental problems created by plastic bags in Ireland. The tax measure was not obvious especially considering the administrative costs related to the tax measure. When even after careful consideration of other measures a tax still seems the best measure to tackle an environmental problem, it is more likely that the right measure is chosen and this prepares the ground for easier implementation of a tax.

When implementing an administratively challenging levy like the plastic bag tax, it is important to carefully consider alternative implementation methods and use existing tax collection methods, in this case the VAT system, to help reduce administrative costs.

6.6. The Norwegian aviation fuel tax

Why was the aviation vehicle tax implemented?

Taxes on air transport are almost non-existent in most countries. Today Norway is one of only a few countries that have aviation fuel taxes. An analysis of the Norwegian aviation fuel tax gives some experiences that other countries might find useful (OECD, 2005g).[30] In 1999 a CO_2 tax on aviation fuel was implemented in Norway as a part of a larger proposal for a Green Tax Reform. Based on recommendations from a Green Tax Commission the Government proposed an increase in green taxes in many different areas. One of the main proposals was to extend the CO_2 tax. It was proposed that almost all end uses of fossil fuels, including aviation should face a minimum CO_2 tax of NOK 100 per tonne CO_2.

How was the aviation fuel tax implemented?

At the same time the Government recognised that the level of taxation would have to be determined on a pragmatic basis, where the declared Norwegian political goal of being a frontrunner in climate change policy would have to be weighed against the costs. The adjustments in the tax should therefore be considered in relation to the level of CO_2 taxes imposed in other countries. To reduce the increased costs for the most affected sectors and regions, the Government proposed that these were compensated. Therefore, when the aviation fuel tax was introduced, the existing seat tax on domestic flights was decreased accordingly. Thus in sum, the aviation sector was not supposed to face increased tax burden as a result of the proposals.

Main results

Because of many other changes in the various taxes and duties on air traffic it is difficult to assess the cost impacts on the airlines due to the fuel tax. However, the total level of aviation charges seems to have increased substantially since 2001 mainly due to changes related to security measures. The aviation fuel tax seems to be of minor importance in this picture because of its small revenue compared to the revenue from other duties. The effects on air ticket prices and thus air travel demand seems to have been negligible due to increased competition in the domestic market and cost reduction programmes imposed by the airlines. Environmental impacts of the tax have probably been negligible. However, the marginal effects of the tax cannot be isolated from the effects of all other changes in the aviation sector over the years.

Main problems/criticisms

The proposals for a tax on aviation fuel encountered opposition mainly from the airlines. The main arguments from the airlines against the aviation fuel tax may be summarised as follows:

- The airlines argued that CO_2 taxes were already high in Norway compared to most other countries and that an introduction of CO_2 taxes in other countries was unlikely. They feared that the aviation fuel tax would prevent Norwegian airlines from competing with EU airlines on equal terms in the forthcoming deregulation of aviation in the European Union.

- Also, the airlines questioned the anticipated positive effects on CO_2 emissions of the aviation fuel tax. They warned that the tax could result in more tanking in neighbouring countries, causing a distortion in competition between airlines with mainly domestic flights and those with extensive international traffic.

- Furthermore, the airlines argued that the aviation fuel tax could be problematic with regard to international agreements.

In May 1999 the airlines were heard on one point: the Parliament asked the Government to appoint a commission to investigate the competitive conditions in air traffic, both nationally and internationally. The Commission appointed consisted of representatives from both the authorities and the aviation sector. In its final report,[31] the Commission concluded that the aviation fuel tax caused competitive distortions between airlines with mainly domestic flights and those with extensive international traffic. The conclusions of the Commission were seen as a victory by the airlines. However, the Government made no changes in the tax as a result of the report.

Main lessons

The aviation fuel tax was introduced without much debate in the Parliament. Indeed there was opposition from the airlines, but they did not manage to get attention in the public debate. ECON points at several reasons for this:

- Most political parties shared the ambition that Norway should continue to play a role of an international pioneer in environmental policy, particularly in imposing green taxes.

- The proposal was only one of several proposals for an environmental tax reform. The tax on aviation fuel was overshadowed by other parts of the comprehensive reform.

- The additional costs incurred by the tax would be compensated by an equivalent reduction in a tax on passenger seats. This made the introduction of the tax far less controversial than it would otherwise have been. ECON points at this as one of the main reasons for why the tax did not meet more opposition.

One of the main arguments regarding competitiveness from the airlines was that airlines with international flights could avoid the tax by tanking abroad. This could have consequences for the competitive position for some airlines as well as for the potential environmental effects of the tax. This argument got some support from the commission that considered the competitiveness issue when they concluded that the aviation fuel tax caused competitive distortions between airlines with mainly domestic flights and those with extensive international traffic.

However, experiences so far seem that the actual possibilities of tanking abroad are limited. There are costs connected to tanking abroad, since the weight of the aircraft increases, resulting in considerable increase in fuel usage which would reduce the net savings. In retrospect ECON concludes that tanking abroad to avoid the fuel tax seems to have been limited, which may also be due to the relatively low level of the tax.

The Norwegian experience demonstrates that fuel taxes on domestic flights are indeed an option. Such taxes should treat all air carriers equally. However, opposition against the tax can be expected. Important lessons from the Norwegian example can be:

- The prospects of tanking abroad should be carefully considered before imposing a fuel tax and also when setting the level of the tax.

- When loss of competitiveness is an issue, compensation, for instance through reduction of other levies, should be considered.

- Countries should strive for the broadest possible tax base in order to ensure cost-efficient emission reduction. Introduction in connection with a broader reform strategy might make it somewhat easier to get acceptance for the tax from affected parties and thus might contribute to a smooth implementation.

6.7. Measures used in practice to limit negative impacts on competitiveness

As shown in some of these *ex post* studies, different measures are used in practice to limit negative impacts on sector competitiveness. It could be useful to sort out the different measures often used and add some comments regarding these measures.

As also demonstrated by the case studies, the use of exemptions and/or rate reductions is common practice to protect (loss of competitiveness) firms that could be strongly affected by the introduction of new environment economic instruments; *e.g.* when Norway introduced the CO_2 tax on aviation fuel the existing seat tax on domestic flights was reduced accordingly and as a result the domestic aviation sector was not supposed to face an increased tax burden. The "Climate change levy" in the United Kingdom offers 80% discount for energy-intensive firms that have signed up to binding negotiated agreements on energy efficiency.[32] As stated in OECD (2001a), other options for mitigation or consumption measures include the introduction of compensation schemes, the use of "border tax adjustments", the recycling (at least part) of revenues from environmentally related taxation, and the co-ordination of countries interested in similar (market-based) approaches.

Exemptions and rebates tend to create inefficiencies in pollution abatement and to undermine application of the polluter pays principle. Indeed, blanket exemptions for polluting products along with rebates for heavy polluting industries can significantly reduce the effectiveness of environmentally related taxes in curbing pollution and similarly reduce incentives for developing and introducing new technologies. A way to get round this problem was sought in the case of the Climate Change Levy in the United Kingdom: an 80% tax rate reduction was offered to sectors that undertook detailed emission reduction targets through the so-called Climate Change Agreements (CCA).[33] As indicated in OECD (2005b), there is, however, considerable uncertainty as regards the "additionally" of the CCAs. There are some claims that the targets set in the agreements represented little more than Business-as-Usual and that the tax rate reductions offered to industry in the Climate Change Levy do indeed reduce the environmental effectiveness of the tax.[34]

Relatively modest compensation mechanisms can often suffice when introducing a tax or a trading scheme (even based on auctioning), in order to make the owners of the firms equally well-off as before; but the size of the "necessary" compensation depends on how insulated the domestic market is from international competition. If some degree of compensation is "the price one has to pay" in order to be able to put in place a new environmental policy measure, this can often be done at a fairly modest cost to society as a whole. However, there is a risk that the affected firms could be seriously over-compensated – in part because policy makers tend to forget (and industry has no incentive telling them) that any environmental policy measure that seeks to limit emissions (and indirectly production) automatically will create a "scarcity rent". In the case of emissions trading, grandfathering of all permits means giving away for free all of this "scarcity rent" – which, based on the findings above, would represent a significant over-compensation.[35] The prime reason why the efficiency costs of providing compensation could be limited is that far from all the "scarcity rent" would have to be given to the most affected firms in order to maintain their profit or equity value. If an "unnecessary" large share of the rent is given to the firms, the economic efficiency costs will increase because less money would be available, for example, to reduce distortionary taxes (cf. the double dividend hypothesis).

Recycling (part of) any tax revenues back to the firms or sectors in questions can be done in ways that secure maintain firms' incentives to abate emissions at the margin, e.g. if full tax rates are paid on emissions or on inputs, as relevant, while a refund is given based on the historic production levels of the firm. Likewise, even if (some of) the emission allowances in a trading scheme are grandfathered, the firms in question will get a full incentive to abate emissions from the alternative value (or "opportunity cost") that each allowance represents.

In fact, when considering introducing compensation schemes, it should be kept in mind that such measures exclude the possibility to use the same revenue for other purposes. Three options are often referred to in the context of environmentally related taxes:

- reductions in taxes on labour, in order to promote economic efficiency and employment;
- measures to compensate low-income households for any tax increases; and
- increases in public spending to protect the environment.

The first option is often referred to in the context of a so-called "double dividend", where increasing environmentally related taxes and reducing labour taxes is believed to bring both environmental improvements and increases in employment. While the first claim is undisputed, the latter is not.[36] There is, however, general agreement that using revenues to reduce labour taxes contributes to higher economic efficiency than if the same revenue *e.g.* was distributed lump-sum to all tax payers, or to a few selected firms. The combining of *e.g.* energy tax increases and reductions of social security contributions have been used in several member countries, including Denmark, Germany, Sweden, and the UK, in part to "sell" the increase in the energy tax and to reduce the tax-wedge on labour.

The second option will be further discussed in the section on distributive issues; however we can summarise here that among the approaches that may be considered to provide relief from an environmentally related tax are:

- an increase in the basic personal allowance (or introduction of an environmentally related tax allowance),
- the introduction of a "wastable tax credit"; and
- the introduction of a "non-wastable tax credit".

The option of increasing public spending to protect environment is again somewhat controversial. While increasing the spending on some environmental purpose simultaneously with *e.g.* an increase in an energy tax can help increase the public acceptance of the latter, there is a significant danger that the "earmarking" of revenues from certain taxes can create rigidities in the budget process and lead to economic inefficiencies.

Finally, regionally or internationally co-ordinated environmentally related taxation would reduce arguments for exemptions and rebates based on international competitiveness; *e.g.* an important step towards gaining acceptance for the Swiss heavy vehicle fee was the bilateral agreement with the EU. However, co-ordinated action does not mean that there will not be any winners or any losers. For example, a global tax on CO_2 emissions would be particularly costly for energy and carbon-intensive economies (OECD, 2001a).

6.8. Political economy lessons from the *ex post* case studies

A first lesson from these *ex post* case studies is that policy-makers should take steps to ensure that competitiveness pressures are adequately assessed and addressed. In doing so, it is important to consider the mitigation measures against any legal obligations and to ensure that the measure will not be found to provide a prohibited subsidy (*e.g.* industrial energy consumption tax in France).

A second lesson is that when loss of competitiveness is an issue, different mitigating measures can be considered and they will have different effects on both environment and competitiveness. When considering different measures it is important that they do not reduce abatement incentives. When levying taxes that raise revenue, many countries have used compensational measures by reducing other taxes (for instance as in case of the Norwegian aviation fuel tax) or other kinds of budgetary compensation. Some countries have introduced sectoral exemptions or reduced rates (as for instance was the case in the UK Climate Change Levy). Finally, sometimes international co-ordination at different levels can be useful and even necessary for implementing market based instruments addressing environmental problems. (as in the case if the Swiss heavy vehicle fee where the bilateral agreement with the EU was important for the implementation.)

However, we should note that there often seems to be a trade-off between the size of the administrative costs and measure to create a "fair" or "politically acceptable" scheme. Often mechanisms introduced for non-environmental reasons, to address competitiveness or income distribution concerns are responsible for the increase of the administrative costs; *e.g.* the CCL in the UK and the MINAS nutrients accounting system in the Netherlands.

Additionally, relatively modest compensation mechanisms can often suffice when introducing a tax or a trading scheme (even based on auctioning), in order to make the owners of the firms equally well-off as before – but the size of the "necessary" compensation depends on how insulated the domestic market is from international competition. However, there is a risk that the affected firms could be seriously over-compensated. If so, the economic efficiency costs will increase because, for example, less money would be available to reduce distortionary taxes.

The "acceptance" of an economic instrument among the public at large seems to be related to the degree of awareness of the environmental problem the instrument is to address. In the case of the Irish plastic bag tax there seemed to be a wide public awareness of the environmental problem caused by littering of plastic bags. The tax therefore seems to have great public support. Therefore, a third policy implication is that it is advisable to "prepare the ground" for later instrument implementation by providing correct and targeted information to the public on the causes and impacts of relevant environmental problems. In general, political acceptance could be strengthened by – as far as possible – creating a common understanding of the problem at hand, its causes, its impacts, and the impacts of possible instruments that could be used to address the problem. One way to build such a common understanding is to involve relevant "stakeholders" in policy formulation, for example through broad formal consultations and/or in committees or working parties preparing new policy instruments. For example, the Swiss heavy duty vehicle tax acceptance was established through referenda and the aviation fuel tax in Norway was seen as a part of a policy shared by most political parties; an ambition to play a role as an international pioneer in environmental policy and particularly green taxes. This acceptance building has been important in many "green tax reforms" in OECD countries over the last decades.[37]

Additionally, the Swiss case is also a good example of the importance of seizing the right moment for pushing through a delicate project on the political agenda. Therefore, a fourth lesson is that a project that at some point in time is impossible to implement might appear to be feasible when the circumstances are more favourable.

A fifth lesson is that countries should strive for broadest possible tax bases to ensure cost-efficient emission reductions. Broad based tax bases and introduction in connection with a broader reform strategy might make it somewhat easier to get acceptance for the tax from affected parties and thus might contribute to a smooth implementation. This strategy also seems to have been followed in many countries that have introduced green tax reforms.

The case study of the Irish plastic bag tax shows the importance of doing thorough initial research and carefully considering other relevant policy options. Introducing a tax is not always the right answer. This study assessed several policy options/instrument to address the environmental problems created by plastic bags in Ireland. The tax measure was not obvious, especially considering the administrative costs related to the tax measure. When, even after careful consideration of other measures, a tax still seems the best measure to tackle an environmental problem, it is more likely that the right measure is chosen and this prepares the ground for easier implementation of a tax. Therefore, a sixth lesson is that, in addition to environmentally related taxes, one should also consider other measures to tackle an environmental problem. This case study also shows that when implementing an administratively challenging levy like the plastic bag tax it is important to carefully consider alternative implementation methods and use existing tax collection methods, in this case the VAT system, to help reduce administrative costs.

Finally, based on the case of the Swiss heavy vehicle fee, one can also draw the lesson that a gradual phasing in of taxes can soften the immediate cost impact and give companies time to adjust to reduce the tax burden.

Notes

1. We should note that all policy instruments can impact in competitiveness, but environmentally related taxes are more "visible" than regulations.

2. The discussion here is based on Delache (2002).

3. With a scale from 1 (hardly open) to 4 (very open), the openness of sectors to international trade in France were ranked as follows: Group 1: printing, press, agro-food, building materials, steel transformation; Group 2: wood, furnishing, paper, rubber, plastics, glass; Group 3: textiles, clothing, mechanical construction, iron minerals, steel, non-iron metals, metals; Group 4: aviation, ship building, leather, shoes, chemistry.

4. Figure 6.1 shows the tax reduction factor corresponding to tonnes oil equivalent per million euro of value added, with a conversion factor 1 EUR = 6.55957 francs. The tax-base reduction factor rises above zero at 25 tonnes (3.81 tonnes) oil equivalent per million francs (EUR of value added).

5. When negotiating a voluntary agreement, it was first necessary to establish a base-case "Business-as-Usual" scenario over a 5-year period, taking into account sector- and firm-specific factors, to be approved by the administration. Then, any annual carbon emission reduction from the base-case scenario would be rewarded, *ex post*, through an annual reimbursement of the tax to the firm.

6. This text is based on a report that Professor David Pearce prepared as part of the work programme of the Joint Meetings of Tax and Environmental Experts under OECD's Fiscal Affairs Committee and Environmental Policy Committee, see OECD (2005b).

7. At that time, the new government faced some potentially difficult climate change goals: committed to reduce carbon emissions by 20% by 2010 relative to 1990 under the Kyoto Protocol; where carbon emissions from the transport sector were rising at the fastest rate, and energy prices were falling because of world economic conditions and market liberation measures (making it more difficult to encourage energy conservation and reduced use).

8. For an in-depth discussion of the political economy of the Climate Change Levy, see OECD (2005b).

9. The CCAs came into being because of the very strong political lobby against the effects of the CCL on the competitiveness of energy-intensive industries.

10. While designed to be revenue-neutral in the aggregate, the reform could generate net receipts or deficits since the social security tax reductions are a fixed percentage (0.3%), and CCL revenues depend on energy consumption by the affected parties and the various exemptions.

11. The combination of the Climate Change Levy and the Climate Change Agreements are discussed further in Section 10.5.3.

12. "Event studies" encompass this kind of analysis but are usually more concerned with effects on company worth, *e.g.* via changes in stock market valuations. Hayler (2003) was unable to detect such an event effect for the CCL, but this is probably due to being unable to capture the complexity of the entire CCL/CCA package in defining the 'event'.

13. The CCAs have interim target dates of 2002, 2004, 2006 and 2008 before the final target date of 2010.

14. A significant part of this reduction is accounted for by the UK steel industry, primarily one company – Corus – which experienced economic downturn conditions in the relevant years. There has been some debate about the extent to which steel's emission/energy reduction is due to the CCA. The CCA for steel does have conditions built into it about renegotiating targets if output falls by more than 10%. See ENDS Report 339, April 2003, pp. 23-26 published by Environmental Data Services Ltd (The ENDS Report, issued monthly, documents virtually all of the claims by individual industries). See *www.endsreport.com/*.

15. Natural gas has 14 000 grams of CO_2 as C per GJ, whilst coal has 26 000.

16. The commitment on the yearly increase of the FDE was 3% in spring 1993, 5% in autumn 1993 and 6% from 1997 until 2000.

17. Households in "fuel poverty" were estimated at 1.75 million in 2002, compared to 4 million in 1996 (UK DTI, 2004).

18. This is true as long as emissions are measured from generation only. On a life-cycle basis, *i.e.* including fuel extraction, station construction, etc., nuclear power is not carbon free. However, even with life cycle effects allowed for, emissions from nuclear power are still small relative to other fuel cycles – see Bates (1995).

19. As quoted in *Accountancy Age*, 25 November 2003.

20. The major part of the CCL revenues are recycled directly back to the corporate sector in the form of social security refunds, and the other revenues partially finance The Carbon Trust. This trust was established in 2001 as a private company and is also financed with grants given by DEFRA.

21. The MINAS is discussed further in OECD (2005d) and OECD (forthcoming).

22. The manure policy in Netherlands consists of three phases (Henkens and Van Keulen, 2001): *i)* stop growing animal production (1984 to 1990), *ii)* step-wise decrease of the manure burden (1990 to 1998), and *iii)* balance inputs of N and P to outputs of N and P (1998 to present).

23. The nutrient accounting system MINAS was implemented at farm level in 1998 on livestock farms with more than 2.5 livestock units per ha, and in 2001 on all other farms. MINAS was discontinued from 1 January 2006, in response to a finding in the European Court of Justice.

24. Dairy farming covers more than 60% of the agricultural land and has a large share in total nutrient surpluses in agriculture and in total nutrient losses to the wider environment. Most dairy farmers have learned more about integrated nutrient management during three years of MINAS than during 10 years or more of using fertilizer recommendations and so-called fertilization balances.

25. A strongly debated change was the exclusion of P fertilizer as an input on the MINAS balance, in response to (political) pressure from arable farmers to block implementation of MINAS in arable farming in 2001, unless P fertilizer was excluded from the accounting.

26. The European Court condemned the first Action Plan of the Netherlands in its arrest released on 2 October 2003. The main arguments given by the European Court were *i)* the MINAS system does not comply with the regulatory system prescribed by the Nitrates Directives, *ii)* the application limits for animal manure and the levy-free surpluses set for the years prior to 2000 were too high, and *iii)* essential regulations of the manure policy were implemented too late

27. Initially, there were some errors in the fixed default values and these had to be adjusted. Also the size of the levy-free surpluses and the size of the levies have been frequently adjusted.

28. This report was written by Ueli Balmer, Federal Office for Spatial Development in Switzerland, and it was prepared as a part of the work programme of the Joint Meeting of Tax and Environmental Experts under OECD's Environmental Policy Committee and Committee on Fiscal Affairs.

29. Should a retailer not implement the levy correctly, he becomes liable to a fine of up to EUR 12.7 million, or to imprisonment for up to 10 years. *One* retailer has thus far been convicted for not charging a customer the EUR 0.15 levy – and, upon pleading guilty, was fined EUR 150, see Lamb and Thompson (2005).

30. Report prepared by the Norwegian consultancy firm ECON Analyse as a part of the work programme of the Joint Meeting of Tax and Environmental Experts under OECD's Environmental Policy Committee and Committee on Fiscal Affairs.

31. See Commission investigating the competitive conditions in air traffic (1999).

32. Some of these exemptions are reviewed in OECD (2001a).

33. Energy-intensive firms in Germany have obtained a 97% reduction in the tax rates of the ecological tax reform there – without any new commitments to specific emission reductions, apart from the general acknowledgement of the environmental agreement on climate change. According to Green Budget Germany (2004), the UK chemical industry pays EUR 0.14 cents per kilowatt hour tax on electricity and in return, agreed to reduce emissions by 18%, the UK aluminium industry pays the same low rate and committed to reduce emissions by 32% in return. In Germany, both sectors are required to pay EUR 0.06 cents per kilowatt hour – without accompanying agreements on company level, but at least a sectoral commitment to reduce GHG emissions.

34. For additional discussion of the environmental effectiveness and economic efficiency of the combination of the Climate Change Levy and the Climate Change Agreements, see OECD (2003c) and Braathen (2005).

35. All the SO_2 emission allowances in the US Acid Rain programme were grandfathered for free. In the new EU CO_2 emissions trading scheme, over 95% of the allowances for the period up to 2007 were grandfathered for free.

36. See Chapter 4 and OECD (2004c) for a further discussion.

37. The issue of acceptance building is discussed in OECD (2001a) and in Chapter 9 below.

ISBN 92-64-02552-9
The Political Economy of Environmentally Related Taxes
© OECD 2006

Chapter 7

Impacts on Income Distribution

7.1. Background and empirical findings

In addition to the issue of loss of competitiveness, the issue of income distribution is often raised when environmentally related taxes are discussed. Policy makers implementing such taxes are often faced with the challenge that some taxes are regressive (progressive) in the way that the budget share of the taxed product tends to be higher (lower) for households with relatively low total household expenditure.[1]

The distributional effects of environmentally related taxes can arise from a variety of channels. The broad categories of distributional effects are:

- The *direct distribution* effects on households arising from payment of the tax.

- The *indirect distributional* effects i.a. from price increases on taxed products from firms.

- The effects arising from the *use of environmental tax revenues*.

- The effects relating to *benefits of environmental improvements*.

Most studies available on income distribution impacts of environmentally related taxes tend to focus on energy/carbon taxation. Available evidence suggests that the direct effect of energy taxes tends to be income regressive. Brännlund and Nordström (2004), Cornwell and Creedy (1997), Symons *et al.* (1994), Tiezzi (2001) and Labandeira and Labeaga (1999) are examples respectively from Sweden, Australia, the United Kingdom, Italy and Spain of household studies. The Australian and Swedish studies confirm the general view that carbon taxes are regressive. Other studies give more mixed results. The Italian and the Spanish studies do not sustain the presumed regressivity of carbon taxes. In their study of the distributional effect of taxes on transportation in Norway, Aasness and Røed Larsen (2002) observe that at least on some environmental indicators the transportation choices made by households with relative high expenditure pollute more given transportation goals than choices made by households with relatively low expenditure. Thus, differentiating the indirect tax system to account for the environmental effects will at the same time reduce inequality. This result is also confirmed in data from the USA by Røed Larsen (2004). The studies are summarised in Table 7.1.[2]

The empirical evidence also indicates that the degree of regressivity decreases once the *indirect effects from price increases for taxed products* are taken into account. This is particularly true for energy taxes and arises from the fact that energy is an input into all goods and services and as such the regressivity associated with the direct effect of tax on a good which is income inelastic is partly attenuated once the effect on all expenditures are taken into account.

Another indirect effect that seems to be analysed to a lesser extent is how environmental taxation can influence firm's demands for capital and labour inputs, which can influence the returns to owners of capital and to labour. Fullerton and Heutel (2004) provide a first theoretical analysis of the incidence and distributional effects of environmental policy that allows for fully general forms of substitution among production factors (labour, capital and pollution) and that solves for all general equilibrium effects of

Table 7.1. **Selected studies on income distribution impacts of environmentally related taxes**

	Authors and date	Main findings
Australia	Cornwell and Creedy (1996)	Carbon taxation – Regressive but transfer payment can be adjusted to compensate for the regressivity without decreasing total revenue.
Italy	Tiezzi (2001)	Presumed regressivity of the carbon taxation is not sustained.
Norway	Aasness and Røed Larsen (2002), Røed Larsen (2004)	Taxes on transportation – Higher taxes on air flights, taxis and automobiles together with lower taxes on bus rides, bicycles and mopeds reduce inequality. Taxes on gasoline contribute somewhat to inequality. Railway passenger transport seems to be neutral distributionally.
Spain	Labandeira and Labeaga (1999)	Neutral effect of carbon taxation across Spanish households
Sweden	Brännlund and Nordström (2004)	Regressivity of carbon taxation. Urban/rural dimension, rural households mostly affected.
United Kingdom	Symons et al.	Carbon taxation – Regressivity depends on how revenue is recycled. Study limited to taxation of driving fuels.
United States	West and Williams (2004)	Increasing the gasoline tax is generally regressive but the results greatly depend on how the revenue is used. Lump-sum may make the gasoline tax regressive.

a pollution tax. They find that a small increase in a pollution tax alters the return to labour relative to capital in a way that depends on the substitutability between the factors. These results provide evidence that the substitutability of capital, labour and emissions has very important consequences for environmental policy and that more work needs to be done to analyse these effects, for instance to calculate the effects of these price changes on different income groups.

The total income distributional effect of environmental taxation also depends to some extent on *how the tax revenues are used*. In many OECD countries revenues from implementation of environmentally related have been used to reduce other taxes for efficiency reasons or to compensate certain groups for income distributional reasons. The total tax package when effects of this use are taken into account may then end up being progressive.[3] This is for instance analysed by West and Williams (2004). They find that while an increase in the gasoline tax generally will be regressive, but this depends greatly on how the revenue is used; using the additional revenue to provide lump-sum transfers more than offsets the regressivity of the gas tax and thus makes the increasing gasoline tax somewhat progressive. They also find that using the additional revenue to lower taxes on labour yields an efficiency gain and makes the policy more progressive, but not enough to overcome the regressivity of the tax. From these results they state the following:

> "Many European countries have already implemented environmental tax reforms that use environmental tax revenues to reduce labor taxes and thus generate efficiency gains in labor markets. Our results suggest that these labor tax cuts will also mitigate the regressive effect of environmentally related taxes. Devoting a portion of the revenue to a lump-sum transfer, or to another progressive policy such as an increase in the earned income tax credit, could even make the net effect of the policy progressive."

Finally, *benefits from environmental improvements* also will have an effect on the total income distribution. Empirical literature generally seems to conclude that low-income households tend to be relatively more exposed to environmental hazards. Especially, a correlation of exposure to environmental hazards with low income is established by several studies.[4] Hamilton (2003) gives an extensive review of empirical work available in North America and other OECD regions. This suggests that the total income distribution can be less regressive when environmental effects are taken into account.

7.2. Ways to address the income distributional impacts

A consequence of this regressivity is that the distributional effects of environmentally related taxes have become a key issue in the policy debate. Experience shows that it is important that the main effects are addressed where necessary, in order to build public support for such taxes. Even if most environmentally related taxes will have a very limited impact on net disposable income of most households, it will be very difficult to "sell" a tax that is *perceived* as "*unfair*". For instance, there seems to be a strong perception in the United Kingdom that taxes on domestic energy use would in particular hurt the "fuel poor" households – *e.g.* low-income households often living in dwellings with very low thermal insulation and outdated heating systems.

When choosing to address income distributional concerns there are several possibilities. Two types of corrective measures can be envisaged to reduce any negative distributional impacts of environmentally related taxation; *mitigation* and *compensation*. Mitigation is an *ex ante* measure to reduce the rates of environmentally related taxes and therefore alleviate the tax burden (both in general and) for specific groups. Compensation measures are basically *ex post* and outside the realm of the taxes as such, *i.e.* they do not affect the rate or structure of the given environmentally related tax.

Mitigation measures can take different forms. It can result in setting tax rates lower than the level that would be optimal[5] from the view of reflecting the marginal social cost of the externalities caused. Such a mitigating strategy can be chosen *i.e.* when introducing new environmentally related taxes to soften the change to a new regime for the involved parties. Mitigating measures can also take forms of introducing more than one tax rate, either a low rate or an exemption (0-rate) for certain groups, *i.a.* low income households. Mitigating measures can also take the form of consumption floors, below which no tax is levied. Introducing mitigating measures of this kind will reduce the environmental efficiency of the tax by removing or reducing the incentives to change consumption and investment behaviour. Mitigating measures are therefore not good practice.

In the case of regressivity governments should seek other and more direct measures that can maintain the price signal of the tax whilst reducing the impact of the tax on household income. One compensation mechanism that retains the incentive effect of the tax is to use lump-sum transfers within the tax and benefit system. In economic terms, these transfers are similar to lump-sum rebate of an environmentally related tax.

7.3. Some considerations regarding different compensational measures

Undesirable distribution effects can in general be addressed through the *social security systems* and *tax systems*. The choice between systems will depend on the country situation, for a number of reasons.

First, the reach of the social security systems, and the reach of the personal tax systems, vary between countries. Welfare lists in certain countries may not cover household groups that are judged to need protection from the effects of environmentally related taxes, whereas in other countries they may. On the tax side, taxpayers with income below a given threshold may not be required to file a tax return, whereas in other countries all residents may be required to file.

Second, the distribution effects of various environmental tax reforms will vary across countries.

A number of approaches may be considered to provide relief from an environmental tax through a country's personal income tax system. These include:

- an increase in the basic personal allowance (or introduction of an environmental tax allowance);
- the introduction of a "wastable tax credit";
- the introduction of a "non-wastable tax credit".

Unlike tax credits, the amount of tax relief from a tax allowance depends on the taxpayer's marginal personal income tax rate. This is because allowances are deductions from the tax base, whereas tax credits provide dollar-for-dollar reductions in tax payable. Where personal tax rates rise with the level of taxable income, a fixed personal tax allowance provides more tax relief to high-income individuals than to low-income individuals, tending to *aggravate rather than alleviate regressivity* in the tax system. Such a result may be avoided by denying a tax allowance to individuals with taxable income above the first positive income band in the rate schedule. However, this may conflict with the intended target group to obtain relief from environmentally related taxes. More generally, on account of interactions with the tax rate schedule, reliance on an environmental tax allowance may constrain policy choices (over the target group, and possibly over the personal income tax rate structure), leading one to consider alternative approaches.

Wastable tax credits are attractive, relative to tax allowances, because they avoid inter-actions with the tax rate structure. However, wastable tax credits do not deliver in full the intended amount of tax relief where an individual has insufficient income (and therefore insufficient tax payable) to fully absorb the tax credit.

Non-wastable tax credits provide *cash transfers* for credit amounts that cannot be used to offset personal income tax liabilities. However, like the other tax relieving measures, they exclude from the net individuals that do not file tax returns. This may not present a major problem in countries where all residents are expected to file tax returns, and may indeed be a desirable feature in countries that aim to discourage non-filing. But where filing tax returns is not expected, other complementary approaches for low-income households may be required.

7.4. Categories of households with special need for compensation

Sometimes when implementing environmentally related taxes some categories of households seem to be in special need for compensation. For instance elderly people may have higher home fuel requirements, as they tend to spend more of the day indoors and may be more vulnerable to colder temperatures. Households with children may have higher heating requirements, tied to the need for a larger living space. One can hence argue that such households should receive more relief from certain environmentally related taxes than others.

A study by Ekins and Dresner (2004) of the UK Policy Studies Institute presents detailed analyses of possibilities to compensate low-income households for new environmentally related taxes or charges in the areas of domestic energy use, passenger transport, water use and waste collection in United Kingdom. They state that:

"In general, it is possible to solve the regressivity problem sometimes associated with environmental taxes and charges through either tariff/charging design or a targeted compensation scheme.

However, the consumption of key environmental resources tends to be widely distributed about the mean within a given income group. This means that, under any practicable compensation system (and assuming no change in household behaviour), some low-income households will end up as net losers from any charging-plus-compensation scheme, even when most low-income households end up as significant gainers."

The problem according to their study is in particular related to domestic energy use. They state that:

"..., the enormously skewed distribution of energy consumption within the income deciles means that the average result conceals great differences in net gains and losses within each decile. In fact, none of the [thirteen] investigated compensation packages manages to reduce the proportion of losing Decile 1 households much below 20 per cent. The conclusion is that, although redistributing the revenues from a carbon tax through means-tested benefits would certainly be progressive overall, and would bring some households out of fuel poverty, no way of effecting such a redistribution was found that would not also worsen fuel poverty for those who are already most badly affected by it. This makes introducing a carbon tax on household energy use politically problematic at best, and probably politically infeasible."

Their findings has been seen to justify the UK Government's opinion that the "fuel poverty" problem has to be "solved" before any tax increase on domestic energy use can be considered.[6]

However, it is important to underline that Ekins and Dresner (2004) looked at the (large) variations in energy consumption across taxable *income* groups. It is probable that *some* of the people that have large energy consumption relative to their income are relatively rich persons with a low taxable income.[7] It is also probable that *a part of* the high energy expenditures is explained by *e.g.* wasteful "heating" practices (leaving the heater on below an open window at winter time, etc.). Finally, it would *a priory* seem possible to design a compensation scheme directly targeted at ("needy") persons with very high energy expenses compared to their income. Hence, the conclusions of Ekins and Dresner can seem too pessimistic.

In their study Ekins and Dresner also point out that households in practice can change their behaviour in response to taxes and charges and reduce resource consumption and thus reduce the number and the extent of net losing low/income households from any tax or tax charging system. This is also pointed out by Schlegelmilch (2003) when considering the distributional effect of the Ecological Tax Reform in Germany. He states that:

"Especially in the field of energy there are numerous possibilities to save that are worthwhile for the individual and could reduce the burden imposed by the ecological tax reform. As a rule it is possible for most households to reduce their burden imposed by the ecological tax by changing simple behavioural patterns."

Since these behavioural changes seldom are included in statistical analyses of distributional effects, the burden imposed by the eco-tax are thus often overstated in statistical terms. Thus, when taking account of behavioural responses the compensation that is necessary to retain the former level of consumption will, in most cases, be smaller than the burden estimated by static models. One should have this in mind when considering designing and sizing detailed measures trying to compensate smaller groups for estimated costs from levying environmentally related taxes.

Further, in order to make the compensational measure efficient, the measures addressed to a special group of individuals should be targeted directly at the factors that cause the equity problems. For instance, although low income is a factor that influences energy efficiency in households, there are other factors, like tenure (renting, not owning) and lack of capital to invest in more energy efficient heating and electrical equipment. Therefore, where the root cause is not low income, other policies, for example direct regulation and subsidies, might be more effective than mitigation and compensation measures.

7.5. Addressing distributional effects in practice – some country examples

When implementing environmentally related taxes that have regressive impact, it is, as pointed out earlier, important that the main effects are addressed where necessary, in order to build public support for such taxes. Otherwise it will be very difficult to "sell" a tax that is *perceived* as *"unfair"*. This has probably been the experience in several OECD countries implementing environmentally related taxes. The box below contains a couple of examples of measures that were not implemented because of income distributional concerns. Compared to a situation with appropriate environmental measures in place, a situation of no implementation will imply higher social costs than necessary.

Box 7.1. **Examples of non-adoption of environmental measures due to income distribution concerns**

In **Ireland**, an inadequate consideration of income distribution concerns contributed to the abolition of domestic water charges during an election campaign in 1996. The water charges operating at that time were based on an unmetered system. This resulted in infrequent and large bills strongly burdening low-income households without standardised methods for dealing with vulnerable families. More recently, the population reacted strongly to the implementation of a new charge for municipal waste services (known as "bin tax") on the grounds that the option of the flat fee retained would be regressive, hitting lower-income families hardest.

In the **United Kingdom**, the government has excluded households from the climate change levy. The reason for its exclusion is related to the government's commitment to address fuel poverty.

Turning to reforms that have been implemented, the distributional effects of environmentally related tax reforms were examined by Bork (2003) in the light of the *German Ecological Tax Reform* implemented since 1998.[8] The reform consisted of a gradual increase from 1999 to 2003 in tax rates imposed on fuels, heating oil, natural gas, and an electricity tax introduced in 1999. The additional revenue raised by the tax reform has been recycled, *inter alia* with an aim to address distributional concerns through the reduction of pension insurance contributions.

The empirical findings show slight distributional impacts of the ecological tax burden which vary according to the category of gross income considered. On average, these effects are weakly regressive. The distributional impacts are depicted in Table 7.2. The income tax reform to a large extent neutralised the burden of the ecological taxes in most cases and the distributional impacts were relatively small.

Table 7.2. **Distributive impacts of the German ecological tax reform**
Estimated impacts for different categories of gross income, 1993

Annual gross income in euros[1]	Households	Changes in the available income as a result of the ...		Net effect of the ecological tax reform	
		Ecological tax	Reduction of pension insurance contributions	Net effect in per cent (%)	Absolute net effect
	Total number	In per cent (%) of the available income			In €/annum
0-7 500	364 114	−0.88	0.03	−0.85	−51
7 500-10 000	929 696	−0.82	0.03	−0.79	−66
10 000-12 500	1 253 067	−0.75	0.04	−0.71	−75
12 500-15 000	1 574 243	−0.74	0.07	−0.67	−85
15 000-17 500	1 577 219	−0.77	0.14	−0.63	−93
17 500-20 000	1 497 397	−0.79	0.16	−0.63	−108
20 000-25 000	2 827 083	−0.87	0.29	−0.58	−120
25 000-30 000	2 683 597	−0.89	0.42	−0.47	−117
30 000-35 000	2 575 889	−0.91	0.60	−0.31	−96
35 000-40 000	2 314 349	−0.91	0.67	−0.24	−88
40 000-45 000	1 974 531	−0.88	0.71	−0.17	−73
45 000-50 000	1 794 961	−0.86	0.73	−0.13	−67
50 000-55 000	1 570 492	−0.83	0.74	−0.09	−58
55 000-62 500	2 072 066	−0.82	0.66	−0.16	−81
62 500-75 000	2 506 797	−0.79	0.58	−0.19	−103
75 000-100 000	3 122 645	−0.66	0.43	−0.23	−152
100 000-250 000	3 702 473	−0.46	0.19	−0.27	−276
> 250 000	1 060 633	−0.15	0.03	−0.12	−376

1. Figures rounded after conversion of DEM to €.
Source: Bork (2003), based on calculations using the Potsdam micro simulation model, in Bach, S. *et al.* (2001).

In 1996 *the Netherlands introduced a regulatory energy tax* (RET) on the use of natural gas and electricity. In later years, the rates of the RET were raised several times. As RET-rates went up, the rate of the first bracket of the personal income tax (PIT) was stepwise reduced by 2.5 percentage points, explicitly to redress in part the distributional impact of the RET. Table 7.3 shows the cumulative impact of the RET-PIT switch over the years 1996-2001 on net disposable income of several socio-economic groups, including the average production worker (APW). At first sight, the numbers in the table suggest that all groups were made worse off. However, during the five-year period under consideration gross wage levels rose continuously and benefit payments were regularly increased. As a consequence, although the isolated impact of the RET-PIT switch was slightly negative, the large majority of Dutch households saw an overall increase of net disposable income between 1995 and 2002. The administrative costs associated with this compensation mechanism were negligible, since the compensation measures were a part of the annual revision of the personal income tax rate structure. This revision takes place anyway, notably because the bracket lengths are fully adjusted for inflation every year.

Table 7.3. **Impact of the RET-PIT switch in the Netherlands**

Cumulative impact on net disposable income of socio-economic groups, 1996-2001

Household type	Net disposable income	RET	PIT	Balance
	2001, in euro	In % of net disposable income		
Workers (couple, 2 kids)				
– Legal minimum wage	14 855	–2.6	1.1	–1.5
– APW	20 870	–1.8	0.9	–0.9
– 2 x APW	33 000	–1.6	0.6	–1.0
Benefit recipient [1] (couple, 2 kids)	13 700	–2.7	1.3	–1.4
Pensioner[1] (couple)	12 210	–2.2	0.9	–1.3

1. At "official" subsistence level.
Source: Ministry of Finance, Netherlands.

The 1998 tax reform in *Denmark* was phased in from 1998 to 2002 and concerned mainly households. The energy taxes and the tax on petrol were increased by 15-25%. The increase in the energy taxes was intended to maintain the real value of the energy taxes. It was introduced at a time when there had been a drop in the oil prices on the world marked.

The 1998 tax reform implied reductions in the personal income taxes for lower and medium incomes. It implied also compensations for pensioners and other recipients of transfer incomes. The total effect of the adjustments in the tax and benefit system was a considerable redistribution in favour of people with lower incomes. All households with small and medium incomes faced a gain in their disposable incomes as a result of the reform. The reform resulted in a revenue loss in 2002 of income taxes equal to DKK 10 billions and a revenue gain from green taxes by DKK 6 billions and DKK 7 billions from property taxes.

After the implementation of the tax reform in 1998 a committee was set up to evaluate possible mitigation measures for some of the green taxes (in particular electricity and water) The committee should evaluate the possibilities of designing a model with a zero-rate band (or tax floor) that would reduce significantly the effective green tax-burden on low income households. The committee found that it would be impossible to design a good mitigation measure that redistributed the tax-burden from households with low consumption to households with a high consumption. The distributional effect was that low-income families with children and households with more persons would benefit from a zero-rate band solution and that households with a single person and single pensioners would loose. Further, the committee found that the compliance costs were considerable, because the electricity companies and the water works did not have the necessary information about the individual families to be able to calculate the tax payment. Most of this information was available at the income tax authorities and the welfare institutions and other public bodies. Therefore, it was recommended not to introduce such a mitigation measure, but to maintain the compensation system, which was adopted in connection with the 1998 tax reform.

Distributional effects of environmental reforms have been addressed by OECD member countries in a number of cases including the mentioned examples from the Netherlands, Germany and Denmark as described above. These experiences show that regressive impacts from implementing environmentally related taxes are often softened by using the revenue to reduce other taxes. Then the tax reductions can be targeted at low-income groups. In other cases the distributional concerns have not been addressed at all, or have come up late in the process and tackled in a more *ad hoc* fashion. This might lead to large opposition and failure to implement effective environmentally measures and implies higher costs to society than necessary.

7.6. How to secure that distributional concerns are addressed in practice?

One possibility of assuring that distributional concerns are properly addressed is to implement such considerations into the decision making process. There are some practical examples of this from member countries. There are different ways of doing this. One way is to adopt specific *institutional arrangements* such as providing a legal basis for addressing distributional aspects or setting up specialised working groups or committees. See Box 7.2 which includes the example of United States and Denmark.

Another way to internalize the considerations of distributional effects can be to develop specific *guidance documents* for policy makers to provide a framework to address distributional issues. Some examples of such guidance documents are presented in Box 7.3.

Box 7.2. **Examples of institutional arrangements to account for distributional concerns**

In the **United States**, distributional issues have been embedded in the policy making through a series of institutional arrangements since the early 1990's. It appears to be one of the only countries where a compulsory legal framework has been adopted to address distributional effects. The US Environmental Protection Agency (EPA) set up an internal working group, in 1990, to study the links between minority and low-income populations and exposure to environmental hazards. In 1992, an "Office of Environmental Justice"* (OEJ) was also established by the EPA to integrate environmental justice concerns into the environmental programs of the agency. A federal advisory committee was then created in 1993, the "National Environmental Justice Advisory Council" (NEJAC), to provide independent recommendations to the US EPA. In addition, the 1994 Executive Order 12 898 established the Interagency Working Group on Environmental Justice (IWG) to enhance coordination between federal agencies, requiring them to formally address issues of environmental hazards in low-income and minority communities and to develop "environmental justice strategies".

Denmark has set up a committee to assess the distributional impacts resulting from the implementation of the green tax reform in 1998 (particularly arising from household electricity consumption) and to evaluate options to address them (OECD, 2002b).

* Originally named the Office of Environmental Equity.

7.7. Conclusions

Most studies show that the direct effects of environmentally related taxes, and especially energy taxes, can have a regressive impact on the income distribution of households. However, empirical analysis indicate that the degree of regressivity decreases once the indirect distributional effects from price increases on taxed products and the environmental effects of the tax are taken into account. Further, when taking account of mitigation or compensation measures the regressive impact of environmentally related taxation can in most cases be softened and even removed. Then the net effect of the

Box 7.3. **Examples of guidance on policy appraisal addressing distributional issues**

At the **European Union** level, the Commission established a new integrated framework for impact assessment with the objective to ensure that social aspects like distributional issues are considered for each policy proposal, together with environmental ad economic impacts. One of the potential social impacts highlighted is the "distributional implications such as effects on the income of particular sectors, groups of consumers or workers, etc.". This extended impact assessment is to be performed for major proposals from 2004 onwards [COM(2002)276].

The **United Kingdom** has formalised central government advice on how to take account of distributional implications in policy appraisal in the new edition of the Treasury Green Book (HM Treasury, 2003). This guidance applies also to the retrospective evaluation of a policy, programme or project and its completion or revision. According to the significance of the distributional incidence across different groups, including income groups, action may be required to modify the policy in question (Davies and Dunn, 2003).

In the **United States**, the guidance documents for incorporating environmental justice considerations into developing environmental impact statements (EIS) or environmental assessment (EA) issued by the Environmental Protection Agency (EPA) are being implemented.

environmental policy can even end up being progressive. Therefore, a full assessment of the income distributional effects of levying environmentally related taxes should also include indirect distributional effects from price increases on taxed products, effects arising from the use of environmental tax revenues and/or compensational measures, and also and the distribution of the environmental benefits resulting from the tax.

Mitigation practices reduce the environmental effectiveness of taxes. In the case of regressivity, governments should seek other, and more direct, measures if impacts on lower-income households are to be alleviated. Such compensation measures can maintain the price signal of the tax whilst reducing the negative impact of the tax on household income.

Undesirable distribution effects can in general be addressed through the *social security systems* and *tax systems*. Relief from an environmental tax through a personal income tax system can *i.a.* include; increases in a basic personal allowance, introduction of non-wastable or wastable tax credits. Wastable tax credits are attractive, relative to tax allowances, because they avoid inter-actions with the tax rate structure. However, wastable tax credits do not deliver in full the intended amount of tax relief where an individual has insufficient income to fully absorb the tax credit. Disregarding any budgetary concerns, non-wastable tax credits might be preferred because they provide *cash transfers* for credit amounts that cannot be used to offset personal income tax liabilities.

Sometimes when implementing environmentally related taxes some categories of households seem to be in special need for compensation. Compensation measures to specific groups of individuals should be targeted directly at the factors that cause the equity problems in order to make the compensational measure efficient.

Experiences from some member countries show that regressive impacts from implementing environmentally related taxes are often softened by using the revenue to reduce other taxes *i.a.* on income. Then the tax reductions can be targeted at lower income groups. In other cases the distributional concerns have not been addressed at all or have come up late in the process and tackled in a more *ad hoc* fashion. This might lead to large opposition and failure to implement effective environmentally measures and implies higher costs to society than necessary.

In order to assure that distributional concerns are properly addressed, member countries should consider introducing measures that implement considerations of distributional concerns into the decision making process. Some countries have therefore introduced specific *institutional arrangements* as for instance specialised working groups or committees. Other countries have developed specific *guidance documents* for policy makers.

Notes

1. Poterba (1991) presents arguments favouring the use of current annual household expenditure, rather than current annual income in the denominator of a measure of budget share allocated to products subject to environmental tax. Annual household expenditure seems to better than current annual income reflect the ability to pay consistent with a lifetime measure of income. See also Pearson (1992) for a discussion of considerations in interpreting distribution analysis that rank households by total expenditure versus total income.

2. This is, however, not an extensive overview of studies of distributional effects of environmentally related taxes.

3. This is for instance discussed by Bork (2003), who concludes that overall the effects of the German ecological tax reform are weakly regressive. Smith (1998), however, made the point that it can be very regressive to use the revenue to lower income taxes.

4. See for instance Mohai and Bryant (1992a), Yandle and Burton (1996) and Hamilton and Viscusi (1999).

5. Optimal tax rates are briefly discussed in Chapter 2.

6. "Fuel poverty" describes a situation whereby a household would need to spend more than 10% of its income on heating in order to obtain an adequate level of warmth.

7. This *could e.g.* be (elderly) people living off a considerable wealth.

8. On the distributional effects of the ecological tax reform, see also Schwermer (2003).

ISBN 92-64-02552-9
The Political Economy of Environmentally Related Taxes
© OECD 2006

Chapter 8

Administrative Costs

It is possible to design a number of economic instruments for environmental policy with relatively low administrative costs, both for public authorities and the affected firms or households. For example, taxes on petroleum products are levied on a limited number of petroleum refineries and depots, and should hence be relatively simple to administer and enforce.[1] For example, the administrative costs of the ecological tax reform in *Germany* are estimated to comprise only 0.13% of the additional revenues raised, according to Deutscher Bundestag (2002). This was said to be very low, for example compared to the administrative costs of the income tax in Germany.

The tax rates of such taxes can be increased, and be closer related to the environmental impacts of the respective products, with very modest (if any) increases in administrative costs, in part because it would normally not be necessary to put in place new collection mechanisms, etc.

OECD (2005a), discussing the distance- and weight-based road fee for heavy goods vehicles in Switzerland, indicates that also for a scheme involving a large number of tax payers can the administrative costs be kept at relatively modest levels.[2] The same has, for example, also been found as regards the levy on plastic bags that was introduced in *Ireland* in 2002, see Box 8.1.

In *Poland*, a comprehensive system of environmental *charges* is in place. All the charges are collected by regional authorities and then distributed mainly to regional funds and to a

---

Box 8.1. The plastic bag levy in Ireland

As described in Section 6.5, a tax on plastic bags was introduced in Ireland in 2002, with a tax rate of EUR 0.15 per bag.

From an administrative point of view, it was originally thought most efficient to levy the tax on producers and importers of plastic bags, thus *inter alia* limiting the number of collection points. However, it was argued from the side of domestic producers of plastic bags that the proposed tax rate would represent some 1 500% of the net-of-tax price of the bags, which could make smuggling become an issue. From an environmental perspective it was also argued that levying the tax at the point of sale could provide a stronger signal to consumers to avoiding plastic bags.

Based *inter alia* on these arguments, the tax is in fact levied at some five thousand points of sale, according to Lamb and Thompson (2005). Each retailer is obliged to pass on the full amount of the levy to their customers – and local authorities enforce that they do so.

To limit the administrative costs of such an approach, the Irish Office of the Revenue Commissioners developed a solution whereby the collection of the tax was integrated into the VAT collection system. This entailed, according to Convery *et al.* (2005), a one-off set-up cost of some EUR 1.2 and annual administrative costs in the order of EUR 350,000. The retailers' extra administrative costs seem to be more than off-set by cost savings in terms of forgone plastic bag purchases and through additional sales of bin liners.

---

national fund. The charges are calculated by the firms affected, and have to be paid to the regional authorities' accounts. Regional authorities are obliged to verify whether the charges are properly calculated and paid in terms. The environmental funds and the National Environmental Fund are independent, non-profit institutions that provide financing for environmental projects, in accordance with priorities established in the national environmental policy, mainly in form of soft loans.

The *total* costs of running the regional funds (employment, rent of offices, etc.) both for the collection of the charges and for the evaluation of environmental projects vary between 0.8 and 4.5% of the revenues raised (on average 1.9%) while the running costs of the national fund is 0.9% of the revenues they receive.[3]

However, many environmentally related taxes (and tradable emissions schemes) do involve a large number of "mechanisms" that tend to increase the administrative costs. It is emphasised that such mechanisms are often introduced for non-environmental reasons, for example to address competitiveness or income distribution concerns.

For instance, the combination of the Climate Change Levy, the Climate Change Agreements and a CO_2 emission trading scheme in the *United Kingdom* has created considerable administrative costs, both for public authorities and for the firms/sectors involved, see OECD (2005b).

Similarly, the combination of the energy tax and energy efficiency agreements in *Denmark* used to entail very significant administrative costs, see OECD (2003a), *inter alia* because the firms involved were obliged to undertake special "energy efficiency audits" prepared by independent experts. Since the year 2000, such audits are, however, no longer required.

The MINAS nutrients accounting system in the *Netherlands* involved very significant administrative costs, *e.g.* related to the complicated calculations of nutrients surpluses at the farm level, see OECD (2005d).[4]

As part of the preparation of the Action Plan for the Aquatic Environment III in *Denmark*, in-depth analyses of several possible tax instruments were carried out – some of them with similar environmental characteristics as MINAS. No such taxes were eventually included in the agreed Action Plan. Larsen (2004) does, however, suggest that their administrative costs could be kept at a much lower level than in the Netherlands, by calculating the nutrient surplus at a sector level, instead of at the farm level. This could be done by *taxing the supply of nitrogen* through feed and commercial fertilisers from *those who sell these products to agriculture*, allowing them to pass on the tax in the price of the products, and by *reimbursing those who purchase from agriculture*, assuming that this reimbursement also will be passed on back to agriculture. Very important administrative advantages could thus be obtained by moving the levying of tax and the reimbursement of taxes away from the farm level, while the environmental effects of the instrument should remain unchanged.

One lesson that can be drawn from the discussion above is that there often seems to be a trade-off between the size of the administrative costs and measures to create a "fair" or "politically acceptable" scheme.

Another important lesson is that "the devil is in the detail": the exact design of a tax can have important repercussions both for the administrative costs and for the environmental effectiveness in practice. It seems likely that new information and communication technologies can provide lower-cost monitoring and enforcement possibilities for certain types of environmentally related taxes, fees or charges in the years ahead, and/or provide cheaper options for payment transfers, etc.

Some countries have looked in detail at the *administrative costs to businesses* stemming from various information requirements in tax legislation – and in some cases in other legislation too. This has *e.g.* been the case in *Denmark,* where the Government has set a target of reducing the total administrative burden placed on the business sector by 25% between 2001 and 2010. According to the Danish Commerce and Companies Agency (2005), the total tax-related administrative burden levied on Danish businesses in 2004 was almost DKK 8 billion (in the order of EUR 1 billion).[5] About a quarter of this administrative burden was estimated to stem from regulations related to indirect taxes. Among these taxes, about half of the burden was found to be related to the customs law (DKK 650 million) and the VAT law (DKK 350 million).

Four environmentally related taxes entail administrative burdens on businesses estimated to be in the order of DKK 20 million or more on a nation-wide basis: the Motor vehicle registration duty (DKK 78.2 million), the Duty on certain retail containers (DKK 34 million), the Motor vehicle weight tax (DKK 21.5 million) and the Road user charge (DKK 19.6 million). As a comparison, these taxes raised DKK 16.8, 0.6, 8.3 and 0.4 *billion* respectively in revenues in 2004.

The Motor vehicle registration duty was found to create an administrative burden for some 1 400 importers and retailers of motor vehicles. A significant share of the burden can seem to be related to the registration of the vehicles as such – regardless of whether a tax was levied on the registration.

The Duty on certain retail containers affects some 4 350 firms producing or using containers. A major part of the administrative burden is caused by the need to provide a detailed description of the packaging used for each product. The OECD/EEA database on instruments used for environmental policy details 24 different tax-bases under this tax, depending on the size and type of material used in the packaging.

A study on the administrative burden placed on businesses in relation to some indirect taxes, based on the "Standard Cost Method", has also been undertaken in *Norway*, see Oxford Research (2005). In absolute terms, the largest costs were found in relation to the Annual tax on vehicles (NOK 29 million, or about EUR 3.7 million), the Tax on consumption of electricity (NOK 17 million), the Import tax on motor vehicles (NOK 8 million) and the Weight-differentiated annual tax on vehicles (NOK 3 million).

In per cent of the amount of revenues raised, the administrative costs for businesses of the Tax on trichloroethane and tetrachloroethane were the highest (1.76%) and the Weight-differentiated annual tax on vehicles (1%). It is not so surprising that the Tax on trichloroethane and tetrachloroethane comes out on top of such a list, as this tax is explicitly meant to change behaviour – not to raise a significant amount of revenue (less than EUR 1 million in 2004).

The administrative costs for businesses of the Excise on petrol was found to be in the order of 0.003% of the amount of revenue raised, while this percentage for the Auto diesel tax was 0.01%. The percentage administrative costs of the sulphur tax (0.59%) and the Product tax on beverage containers (0.45) were both *relatively* high, while the figure for the Tax on final treatment of waste (0.07%) was relatively low. As a comparison, the similar figure for the VAT tax has been estimated to 0.66%.

A similar study has also been made in *Sweden*, in relation to selected taxes, see NUTEK (2005).[6] Two environmentally related tax laws were included in the study, the Tax on waste and the Energy tax law.[7]

Some 250 firms are affected by the Tax on waste, which raised about SEK 750 million (about EUR 75 million) in revenues in 2004. The administrative costs were estimated to SEK 1.9 million – or 0.25% of the revenues raised.[8]

About 1 150 firms are directly affected by some of the information requirements related to the Energy tax law, and their administrative costs in complying with the information requirements are estimated to SEK 22.6 million. With total energy tax revenues in the order of SEK 55 billion, the cost-to-revenue ratio is in the order of 0.04%.

Notes

1. The very strong emphasis on taxes on motor vehicle fuels in Turkey can probably in part be explained by the administrative advantages of such taxes.

2. A more "advanced" road pricing scheme for heavy vehicles in Germany did run into considerable practical implementation problems, and was implemented with a significant delay in 2005.

3. The fact that these numbers include the running of the financing activities of the funds means that they are not directly comparable to *e.g.* the numbers provided as regards the German tax reform.

4. The costs of calculating the nutrient surpluses at farm level *did* have an environmental motivation. OECD (Forthcoming) does, however, illustrate that MINAS was combined with several other policy instruments that tended to create considerable administrative costs – without much improvement in the overall environmental performance of the instrument mix.
   The MINAS system was discontinued from 1 January 2006 – in response to a judgement of the European Court of Justice that found the Netherlands not to comply with the EU Nitrates Directive.

5. The study was based on the "Standard Cost Method", developed in the Netherlands in the 1990s. The focus is on firm's costs related to retrieve, document, store, make available or report all sorts of information in order to comply with obligations set out in legal texts. Based on in-depth interviews, estimates are made of the amount of time it would take a "normally efficient" firm to comply with the various obligations. To arrive at nation-wide estimates, the number of firms affected (to a varying degree) by a certain regulation has to be taken into account – and a cost per hour spent has to be imputed.

6. NUTEK (2005) contains very detailed estimates of the costs related to all the different information requirements of the laws in question.

7. The Energy tax law covers several specific taxes detailed in the OECD/EEA database on instruments used for environmental policy.

8. The cost-to-revenue ratio in Sweden for the waste tax is seemingly somewhat higher than in Norway, but – even if both studies build on the "Standard Cost Method" – no attempt has been made here to verify if the numbers can be compared directly.

ISBN 92-64-02552-9
The Political Economy of Environmentally Related Taxes
© OECD 2006

Chapter 9

Enhancing Public Acceptance

The "acceptance" of an environmentally related tax among the public at large seems to be related to the degree of awareness of the environmental problem the instrument is to address. For example, littering of the environment by plastic bags was a much-focussed issue in public debate in Ireland prior to the introduction of the levy on plastic bags. In Switzerland there had for many years been significant focus on the nuisance caused by cross-alpine heavy-vehicle transport, prior to the introduction of the distance-based road fee for heavy-goods vehicles in 2001.

These environmental problems were directly, and immediately, perceived by a large share of the population. It *can* be more difficult to obtain acceptance for instruments that address problems where the impacts are not so directly felt by the public at large – like climate change, ozone layer destruction or eutrophication of distant waters.[1]

The so-called PETRAS project studied the attitudes of business and the general public towards environmental tax reform – combining increases in environmentally related taxes with reductions in *e.g.* taxes on labour – in five EU member states (Denmark, France, Germany, Ireland and the United Kingdom).[2] The project showed that such reforms can face a number of problems: there is not so much outright hostility to environmentally related taxes as conceptual problems with their design. Similar conceptual problems were also found in the interviews with business people.

According to the findings of that project, people did not *trust* assurances that the revenues will be used as promised by the government. They also had difficulty in *understanding* the purpose of increasing taxes on energy while lowering taxes on employment. There were also problems with the *visibility* of reductions in other taxes. For example, in Germany and Denmark people were aware of paying higher energy taxes, but few were aware of the lowering of social insurance taxes in those countries. A desire for measures seen as *incentives* as well as penalties was also expressed.

An implication is that it is advisable to "prepare the ground" for later instrument implementation by providing *correct* and *targeted* information to the public on the causes and impacts of relevant environmental problems. Any "scaremongering" should be avoided, as that would tend to lower the public credibility also of correct information provided at a later stage.

Further, the purpose of a new tax, or a tax rate increase, should be made clear from the outset, in particular when the objective is to reduce a specific pollution problem, rather than provide government revenue. This was, for example, done in relation with the Plastic Bag Levy in Ireland (see Box 8.1): extensive multi-media advertising made it clear that the purpose of the tax was to help avoiding that discarded plastic bags spoil the Irish landscape.

Environmental taxes face opposition from stakeholders for a variety of reasons, such as fear of competitiveness losses, reduced profits or possible income regressivity. It seems clear that the degree of political acceptance depend on the perceived "fairness" of the instrument in question. A lot of the attention concerning "fairness" focus on the perceived sectoral competitiveness impacts and/or negative impacts on low-income households. Policy makers could do well in trying to shift the discussion of "fairness" more towards "who are the most important contributors to the problem at hand" – and to try to make people more aware that *any* instrument that could be applied to address a given problem will have (positive or negative) income distribution impacts.[3]

The public acceptance for various "positive" tax measures – *i.e.* preferential tax treatment of the least harmful or most benign alternatives from an environmental point of view – is generally relatively high. Tax rate differentiation between high- and low-sulphur motor fuels, or between leaded and unleaded petrol, are examples of such measures – that seem to have been well understood and received by the public. Another example of such measures is a so-called "feebate" system, where *e.g.* a tax is paid on vehicles with low fuel efficiency, whereas a subsidy is given for fuel-efficient vehicles. A drawback with providing subsidies for the environmentally least harmful option is, however, that the total car stock could thus (marginally) increase. The cost of financing any subsidies is also an issue – the revenues raised through taxes on the most harmful options could, alternatively, for instance have been used to lower distortionary taxes elsewhere in the economy.

This point would, of course, be of even higher importance for any "positive" measure that only included a reduced taxation of a "benign" option, without any increased taxation of the more harmful varieties. Examples of such measures include accelerated depreciation for environmentally preferable capital equipment, tax credits for certain environmental expenditures, etc.

In general, political acceptance could be strengthened by – as far as possible – creating a common understanding of the problem at hand, its causes, its impacts, and the impacts of possible instruments that could be used to address the problem. *Correct* and *targeted* information to the public – *inter alia* through advertising, printed publications and on-line databases – is one type of policy tools that can be used to promote this. Another way to build a common understanding is to involve relevant sector ministries and other "stakeholders" in policy formulation, for example through broad formal consultations and/ or in committees or working parties preparing new policy instruments.[4]

Such commissions may include different government ministries (*e.g.*, finance, environment, energy, industry, agriculture), representatives from the economic sectors concerned (agriculture, energy, transport, industry, etc.), environmental NGOs and technical experts. These "green tax commissions" can provide public and technical legitimacy to the tax reform. They usually have a mandate of several years, enabling them to achieve solid work and progressively build confidence and dialogue. When a tax reform is decided, such commissions may also have a role to play in monitoring and assessing the implementation.[5]

Public acceptance *might* also be enhanced if a reform is implemented gradually. In particular, the initial introduction of environmental taxes can be followed by a gradual increase in tax rates, a widening of the application of the tax and a progressive introduction of new taxes. For instance, in Finland, the rate of the CO_2 tax was raised from FIM 24.5 per tonne of carbon in 1990 to FIM 374 per tonne in 1998; the tax was initially limited to heat and electricity production and later broadened to transport and heating fuels. In Denmark, the CO_2 tax introduced in 1992 was followed by a multi-year "Energy Package" (1995-2002) comprising a progressive increase of the tax.

The so-called "ecological tax reform" initiated in 1999 in Germany included significant annual increases in several of the tax rates. Innovative advertising campaigns in various media were used to increase the publics understanding of the motives behind the reform, and of the impacts that could be expected.

In the United Kingdom, the tax rate for "active waste" in the Landfill tax is now gradually being increased over several years – well beyond the estimated externalities related to landfilling. Public acceptance is, however, probably not so much an issue here, as the public at large hardly sees this tax.[6]

One argument that could be used against a gradual phase-in of a tax increase – as opposed to a one-off larger increase – is that with a gradual approach, a new political "fight" about the tax rate will have to be made each year. It is conceivable – but not at all given – that such ever-returning debates by themselves would tend to increase the opposition against the reform.

As discussed further in Section 10.6, if the use of some form of "carrot" next to the "stick" (i.e. the tax) is seen as necessary in order to obtain public support for a new tax, it could be useful to propose increased spending for a "popular" purpose simultaneously with the proposal of the tax – without making a legally binding link between the tax revenues and future expenditures.

As referred to above, one problem in relation to public acceptance of (new) environmentally related taxes is that many people forget soon any "compensation" given in the form of reductions in other taxes. That is to say, a few years after – for example – a new energy tax was introduced, many people will have forgotten any reduction made – for instance – in income taxes or social security contribution.

While it is not defined as an environmentally related tax, Canadian authorities faced a similar issue in connexion with the introduction of a Goods and Services Tax (GST) some years ago. They chose to introduce a "visible" tax relief provided through the personal tax system, in the form of a "Goods and Services Tax credit". The GST credit was put in place to ensure that families with annual incomes below CAD 30 000 would be made better off under the new sales tax regime. The GST credit is non-wastable,[7] and depends on family size and income. For a married couple with 2 children, the full credit is CAD 650, an amount that is reduced when family net income exceeds CAD 27 749 and fully eliminated when family net income reaches CAD 40 749.

Taxpayers are made aware of the GST credit on several occasions each year. First, to apply for the tax credit, residents are required to complete an annual personal income tax return. The GST credit is calculated in a separate section of the tax return, tending to draw attention to this tax provision. Second, the credit amount is distributed in four payments over the year. Third, information is widely available to the public on the tax crediting provisions. While no empirical evidence is available, it would appear that these efforts have been successful in terms of building and maintaining public support for the GST.

Notes

1. Thalmann (2004) discusses the reasons for why Swiss voters in 2000 rejected three proposals for taxes to limit CO_2 emissions. He found that few voters paid attention to the fine differences between the proposals made. Those who did favoured the lowest tax rates with revenues earmarked for a wide range of subsidies. The promise of a favourable direct impact on employment made by a mini "green tax reform" was not understood or valued.

2. See PETRAS (2002) for more information. For a recent discussion of social and political responses to environmental tax reform in several European countries, see Energy Policy (2006).

3. To just give one example, a tightening of buildings' thermal quality standards *could* deprive low-income households of possibilities for relatively low-cost housing, where they (in principle) could compensate for low indoor temperatures by putting on additional clothes.

4. For a further discussion on the involvement of "stakeholders" in an environmental fiscal reform – with a focus in particular on such reforms in developing countries – see OECD (2005e).

5. Subsequent to a recommendation in OECD (2005h), France has recently established a "Green Tax Commission".

6. For the households, the costs related to this tax are "hidden" as part of the Council tax. The tax rates are set to increase GBP 3 per year until a level of GBP 35 per tonne active waste is reached – as an instrument to help secure fulfilment of the United Kingdom's obligations under EU's Landfill directive.

7. See Section 7.3 for explanation.

—

ISBN 92-64-02552-9
The Political Economy of Environmentally Related Taxes
© OECD 2006

Chapter 10

Environmentally Related Taxes Used in Instrument Mixes

10.1. Introduction

Environmentally related taxes are seldom used in isolation to address a particular environmental problem. Most often they are used in combination with one or more other instrument categories, like direct regulations, subsidies, labelling systems, negotiated agreements, etc. OECD has for some time been studying the impacts on environmental effectiveness and economic efficiency of the fact that such instrument mixes are employed – instead of only one instrument being used per target set. Central to the work is how the different instruments interact in practice. This chapter presents some examples of instrument mixes used and highlights some of the main findings made in this work.[1]

There is a saying about using one stone to kill two (several) birds. Applied to policy making, this saying hints at the possibility of pursuing several policy targets with a single instrument. Textbooks in economics have tended to be sceptical about the possibilities of realising the promises of this allegory in practical policy making – since the publication of Jan Tinbergen's book "On the Theory of Economic Policy" in 1952. He emphasised the *need for having as many instruments as one has policy objectives.*[2]

Conceivably, in using several instruments at the same time, one instrument could help underpin the functioning of another instrument. On the other hand, using more than one instrument to pursue a single target could imply redundancies – with *e.g.* unnecessary administrative costs – and there is also a risk that the introduction of an additional instrument could harm the workings of the instrument(s) already applied.

10.2. Taxes used in combination with regulatory instruments

A large share of all environmentally related taxes are used in combination with one or more (environmentally motivated) regulatory instruments, as some type of regulatory instrument(s) are used to address most environmental issues. A few examples are mentioned below.

A number of countries[3] apply taxes on the sulphur content of fuels, and many of the same – and other – countries apply tax rate differentiation to promote low sulphur content in motor fuels (see Figure 2.7). These tax instruments are in most – if not all – cases combined with regulations that place an upper limit on the sulphur content of the fuels. A *priori*, it does not seem likely that there are strong interactions between these instruments, even if the regulations could place *some* limits on the flexibility normally offered by a tax instrument. A possible rationale for the use of both instruments is that a tax on the sulphur content on (all) fuels can be an effective and efficient instrument to address *total* sulphur emissions within a larger geographical area – that contribute to acidification problems, etc. – whereas the regulations might be better suited to address any *local* air pollution "hot spots", linked to human health problems, etc.[4]

Taxes on the landfilling of waste are always combined with some regulations placing limits on the amounts of certain types of waste that may be landfilled. *For example*, according to the EU Landfill Directive, in most EU member states the amounts of biodegradable municipal waste going to landfills in 2006, 2009 and 2016 are not allowed to exceed 75%, 50% and 35% respectively of the amounts that were landfilled in 1995 in each country.[5] Some countries, like the Netherlands, are simply banning the landfilling of any combustible waste at all.

The rationale for such an instrument combination is not obvious. *If* the landfill taxes are set at an appropriate level to internalise all the negative externalities associated with landfilling – and *if* the taxes are passed on to the collection charges households and other waste generators are facing *at the margin* – there would be few economic arguments in favour of *also* placing specified limits on the amounts that could be landfilled.

However, regulations could usefully be combined with economic instruments in situations were for some reason such instruments do not have significant impact or are practically difficult to implement. For instance, taxes on the landfilling of waste do not give incentives to reduce emissions from waste that is already deposited. Hence, regulations that reduce for instance leakage to groundwater and emissions of methane gas can clearly be useful.

More generally, while taxes often are well suited to address the total use of a given product or service – or the choice between different product varieties with different environmental characteristics – they can, *inter alia* due to monitoring and enforcement problems, be less suited to address *how* a given product is used, *where* it is used, *when* it is used, etc. To the extent that the final environmental outcome also depends on such characteristics – and to the extent compliance with a regulation can be more easily monitored and enforced – it can be useful to combine a tax and a regulation.

There is, on the other hand, always a *danger* that regulatory instruments can *unnecessarily restrain the flexibility* that an environmentally related tax (or a tradable permits system) could offer to the polluters to find the lowest-cost options to address the problem at hand. This would be an example of negative interactions between instruments used in combination. Hence, whenever an environmentally related tax is being prepared, one should carefully consider whether parts of existing regulations have become redundant.

10.3. Taxes used in combination with tradable permits systems[6]

A joint application of tradable permits and a marginal tax, that is, a tax per emission unit such as a carbon dioxide tax, will not contribute to any additional environmental benefits as long as the demand for permits exceeds the supply at the chosen tax rate level. As total emissions equal the number of allocated permits, a marginal tax will only affect the distribution of emission reductions within the trading system. There are in that case no environmental reasons for using taxes in combination with a tradable permits system.

There are, however, other reasons for a joint application of tradable permits and environmentally related taxes. Potential motivations for the introduction of taxes to supplement a trading system are:

- as a means to reduce compliance cost uncertainty;
- as a means to penalise non-compliance; and,
- as a means to capture windfall rents from the permit allocation.

While the second motivation (penalties for non-compliance) is not strictly a tax, the close link between it and the other two cases means that it is relevant to include it in the discussion. These three different motivations – which are not mutually exclusive – will be discussed in turn.

10.3.1. *Using taxes to reduce compliance cost uncertainty*

The potential desirability of the joint application of taxes and permits (rather than using one or the other on its own) has been recognised for many years. In particular, Roberts and Spence (1976) proved that the joint application of the two instruments was preferable in the presence of:

● non-linear environmental damages; and,

● uncertainty concerning abatement costs.

Since both of these conditions are likely to hold in a large number of cases, the practical importance of this result can hardly be overstated.

The point can be illustrated with reference to Figure 10.1. Assume that a firm initially emits 20 units of pollution. If environmental damages rise steeply, the welfare implications of under- or over-estimating abatement costs can be considerable. Assume that a tax is introduced at a level at which the forecasted marginal abatement costs equal marginal damages. The tax would be approximately USD 600 per unit emitted with emissions equal to 7 units (20 minus 13). Alternatively 7 permits could be issued, resulting in a permit price equal to the tax.

If, on the one hand, the marginal abatement cost proves to be higher than assumed, abatement levels will be too high under the permit regime and too low under the tax. If the marginal abatement cost turns out to be lower than expected, too much would be abated under a tax regime and too little under the permit regime. Which results in greater welfare losses? Roberts and Spence (1976) showed that if marginal damages from emissions rise sharply and non-linearly, then the permit regime will be preferable.

In Figure 10.1, if the marginal cost of abatement is higher than expected, welfare losses are greater with a tax (area A) than with a permit system (area B). If costs are lower

Figure 10.1. **The welfare costs of abatement cost uncertainty under permits and taxes**

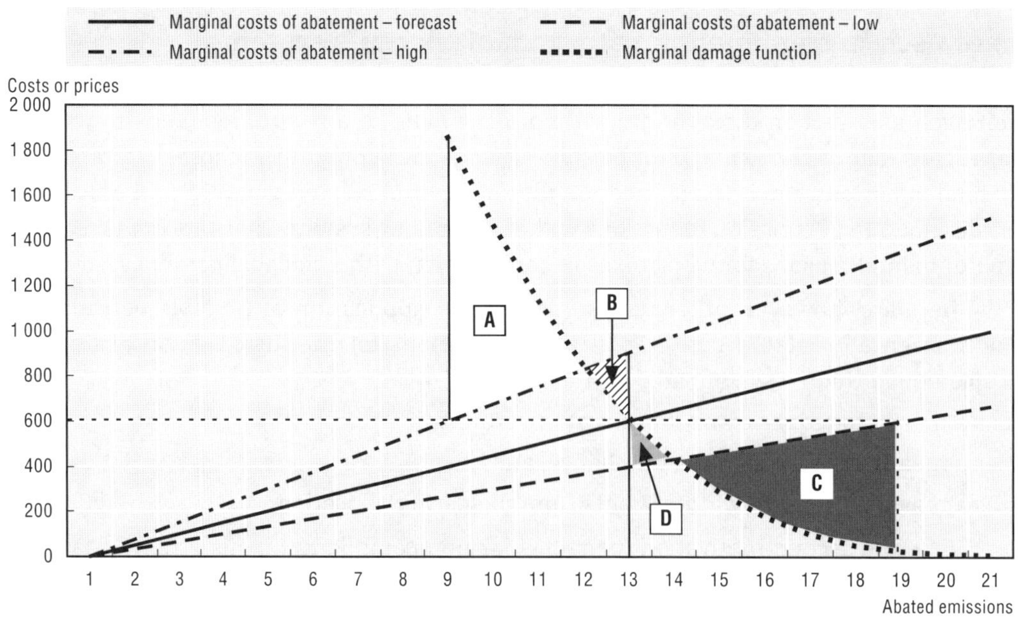

Source: OECD (2003b).

than expected this is also true – area C is greater than area D. With steeply rising marginal damages from emissions, the benefits from having certainty with respect to the level of damages exceed the costs associated with uncertainty about abatement costs. Of course, if marginal abatement costs rise steeply (while being uncertain), but marginal damages from emissions are low, the opposite is true, and a tax is preferable.[7]

However, Roberts and Spence (1976) made the additional point that *a mixed regime is better still*. In effect, by delimiting the bounds of permit price uncertainty through taxes and subsidies, the potential welfare losses from the regulatory authority either over-estimating or under-estimating marginal abatement costs can be reduced. Implicitly, many policy makers have recognised this by using emission taxes as a means to "cap" potential permit prices. The analogue – using subsidies to put a floor on permit prices – has not been used explicitly.

This has been dubbed the "safety valve" argument.[8] By putting a cap on permit prices, regulatory authorities can be able to convince risk-averse affected firms and households of the desirability of introducing a tradable permit regime. In Denmark, the government explicitly used a "safety valve" argument in setting the penalty at DKK 40 (EUR 5.4) per tonne of CO_2 in the previous emission trading system for electricity generating firms.

In the original proposals for the EU Directive on greenhouse gas emissions trading, the permit price cap was set at EUR 50 per tonne in the first phase (up to 2008) and EUR 100 per tonne in the second phase for each carbon dioxide equivalent emitted by that installation for which the operator has not surrendered allowances (CEC, 2001a). However, the original proposal also included the provision that the penalty should be *twice the average market price* if the market price should be higher than these levels. In a subsequent amendment to the Directive this provision was removed (CEC, 2002b)[9]. In effect, by removing this clause the penalty has become a price cap, which would have not been the case under the previous system proposed. Indeed, the Commission explicitly stated that price certainty was the objective of the amendment.[10]

In the United Kingdom the Landfill Tax is combined with a Landfill Allowance Trading System for biodegradable municipal waste. The tax does, however, *not* function as a "safety valve" as concerns the prices of the landfill allowances in this case. On the contrary, the trading system includes it own significant fines that would apply in cases of non-compliance, that would have to be paid on top of the Landfill Tax for any "excess" amounts of landfilling.[11]

Arguably an increased cost certainty can also be achieved with the permit trading programme by keeping permit reserves available for use. For instance, under the US SO_2 Allowance Trading Program, the government initially held reserves of permits which it could have released onto the market if the price had reached USD 1 500 (see Tietenberg, 1998). In practice, permit prices never approached this threshold, but certain initial estimates were sufficiently high to elicit some concern about compliance costs. However, such a scheme has the disadvantage that the price can only be capped for as long as the reserve holds – excessive demand will eventually drive the price higher. Thus, the price effects are less certain, undermining the benefits in terms of reduced cost uncertainty. On the other hand, of course, the environmental effects are more certain with a permit reserve, since under a tax-based price cap the government has no direct control over total emissions.

10.3.2. Using taxes to penalise non-compliance

Financial payments can also be used to penalise non-compliance. In this case, the logic is quite different. Rather than serving as a "safety valve", the tax is designed to serve as a deterrent. In one case (the cap on permit prices) the tax is designed to allow for legitimate and strategic behaviour on the part of the firm. In the other case (the penalty) it is designed to serve as a deterrent on behaviour which is considered to be malign and illegitimate. As such, the penalty would be set much higher relative to the prevailing or expected permit price than would be the case with a tax which is designed to serve as a price cap. For example, the penalty under the Ozone Depletion Substances tradable permit programme in United States was, USD 25 000 per kg.

The optimal level of the penalty will differ depending upon the cost of monitoring. For instance, under the American SO_2 programme there is real-time monitoring of emissions. This means that the probability of being caught for non-compliance is close to 100%.[12] However, if monitoring is imperfect then the penalty will have to be correspondingly higher to serve as an effective deterrent. There is a negative correlation between the probability of being caught for non-compliance and the optimal level of the penalty.

In addition, issues of fairness need to be addressed. For instance, under most systems penalties need to be "reasonable" and "proportionate" to the magnitude of the offence. While it may be possible to improve compliance rates by increasing penalties to "unreasonable" levels, this would be inconsistent with principles of fairness in most legal systems. The penalty level must be commensurate with the nature of the violation.[13]

Since they serve different functions single charges can not serve as both effective deterrents and caps on permit prices. (See Table 10.1 for a comparison of penalties and permit prices under various systems.) The two functions are not complementary. In practice, however, the distinction between them can be ambiguous. Due to abatement cost uncertainty resulting in lower permit prices than anticipated, a penalty can end up serving as more of a deterrent than a cap, as might have been the original intention. This may well have been the case with the American SO_2 Allowance Program where permit prices ended up being much lower than anticipated. For instance, the EPA had made *ex ante* price estimates of over USD 1 000 per ton, but actual prices had converged to less than USD 200 per ton by the mid-1990s. Analogously, if permit prices end up being higher than anticipated, then the tax might be less meaningful as a deterrent, but serve as more of a price cap.

However, it is not just the relative size of the payment which distinguishes the tax and penalty functions. In some cases, the penalty is purposefully left uncertain. For instance, under the UK's Packaging Recovery Notes scheme penalties for non-compliance can be as high as GBP 20 000, however in practice they have only averaged approximately GBP 3 250 (see Salmons 2001). Questions such as intentionality are used in determining the level of the penalty. Similarly, under the RECLAIM programme in California penalties are "discretionary" (subject to a maximum), dependent upon the reason for the violation and subject to appeal. This *ex ante* uncertainty means that even if the penalties are not far in excess of potential permit prices, they are unlikely to serve as an effective cap – or rather will do so in an unpredictable manner.

And finally, since many penalties include additional sanctions – such as the need to purchase additional permits in subsequent years in a proportion greater than one – the penalty will not be an effective indicator of permit price maxima. Firms will need to account for all sanctions (and not just the direct financial penalty) when deciding how many internal resources should go towards ensuring that the firm is continuously in compliance. This may include a number of elements which are difficult for the firm to quantify, such as the "stigma" associated with being in non-compliance.

Table 10.1. **Permit prices and penalties for selected tradable permit systems**

	Permit prices	Penalty
United States – Acid rain	125-225 USD per ton in 2000-2004 (*www.epa.gov/ airmarkets/trading/so2market/alprices.html*)	2 000 USD per ton[2] (*www.epa.gov/airmarkets/arp/regs/sec411.html*)
United States – ODS	n.a.	25 000 USD per kg (Harrison 1999)
United States – NO$_x$	2-3 000 USD per tonne in 2004.	Three allowances for each excess ton (*www.epa.gov/airtrends/2005/ozonenbp.pdf*)
Denmark – CO$_2$	n.a.	40 DKK per tonne (Kitamori, 2002)
EU CO$_2$ Programme	20-25 EUR per tonne, Summer 2005[1] (*www.theipe.com*)	40 EUR per tonne 2005-2007, later 100 EUR per tonne (EU 2003a)

1. Market liquidity and carbon prices have thus far been impacted because not all registries in new member countries of the EU are operational and because the international transaction log is not yet functioning, thereby limiting the introduction of credits based on the "Clean Development Mechanism" – and hence limiting the supply of allowances. As of June 2005, the coal-to-gas spread in power generation also contributed to the EU allowance price of approximately EUR 25 per tonne CO$_2$. At this price, gas-based generation would theoretically surpass coal-based generation on the power market. Any increase in gas prices is immediately followed by an increase in EU allowance prices. However, in 2006 permit prices have decreased significantly, reflecting an over-allocation of permits at the outset.
2. Plus one allowance reduction in the following year's allocation.
Source: Partly based on OECD (2003b).

10.3.3. *Using taxes to capture windfall rents*

The third use of taxes in combination with tradable permit regimes arises from the common use of gratis allocations of tradable permits, rather than auctions. Whether the allocation is done on the basis of historical emissions or regulatory requirements or some other mechanism, firms receive a windfall rent equal to the value of the permits allocated. In order to recover some of these windfall rents, taxes can be applied in conjunction with the tradable permit regime.

This appears to have been the motivation behind the use of the US tax on ozone-depleting substances in conjunction with the ODS permit trading programme. Initially set at USD 1.37 per pound in 1990, the tax rate rose to USD 5.35 in 1995. This tax is paid on all ODS sold and on any stocks of ODS (see Harrison 1999 and section 5.5.3.3 above for more information on the tax). As such, it is *complementary* with the ODS permit trading programme and not a substitute – as is the case with taxes which serve as permit price caps. Thus, irrespective of the permit price, the tax has to be paid. Moreover, it is also complementary to the penalty (USD 25 000 per kg) since that is charged for non-compliance, while the CFC tax is charged on all sales.

Why was this tax introduced in the ODS programme and not elsewhere? Part of the reason is clearly political, with the likelihood of there being considerable resistance to its application.[14] Moreover, the issue of *rent capture* was particularly important, since under the ODS programme firms were effectively receiving a right to produce or import a commodity (see Stavins 2001).[15] The commodity was a permit to produce or import valuable commercial commodities, granted *gratis* to 28 firms. The rent (or windfall profit) arose from the distribution of a production right. This is quite different from being granted permits for a pollutant which can be substituted to a greater or lesser extent in the production of other commodities.

The way the windfall rent is captured will affect the incentives and therefore also the economic performance of the emitters. A marginal tax increases the marginal valve of reducing or increasing emissions equal to the tax. If the sum of the permit price and the tax exceeds the negative externality, this results in a dead-weight loss to society. In contrast, a tax on *e.g.* historical emissions does not affect behaviour at the margin; it only affects entry and exit in the trading system. Thus, in order to avoid distortions of marginal behaviour within the trading system, it is important that windfall rents are capture through such taxes.

10.3.4. Conclusions on combinations of taxes and tradable permit systems

While the intention and motivation for the use of taxes as permit price caps and penalties as legal deterrents are often quite different, in practice they can often have similar effects. This is because if the penalty is set too low, firms will see it as a feasible "compliance" strategy. Similarly, if the tax is too high, it will serve as a deterrent, perhaps encouraging greater vigilance but not being seen as an economic option.

In any event, in order for a price cap to be efficient its use should be explicit, and its size known *ex ante*. A penalty which serves as a default price cap is unlikely to be efficient since penalties are often of uncertain size – for the reasons discussed above. This will have the effect of introducing uncertainty into the market, precisely the opposite effect as the usual motivation for the introduction of a tax.

In some cases it may also be advisable to introduce taxes on the windfall rents associated with the gratis allocation of permits. This is likely to be most important when the permits relate to commercial products (such as CFC's) and not pollutants *per se*. However, this may also be important under certain emission-based permit regimes, depending upon how important the rents are in relation to total compliance costs. This in turn will depend upon the slope of the cost curve and the degree of stringency of the environmental target, *i.e.* the tightness of the cap.[16]

10.4. Taxes used in combination with labelling systems

Due to the existence of numberous information failures, combining a tax with information-based instruments, such as various forms of labels and certification systems, can make both instruments more effective. The labels provide information to the relevant decision-makers (*e.g.* consumers, farmers, industrials) so that they can make better informed choices. Improvements in the quality of the information available to the consumers can help *increase the relevant price elasticities*.[17]

For instance, the provision of information to the consumers can be a case for policy intervention to improve residential energy efficiency. The relatively slow adoption of environmentally preferable consumer durables may to a large extent be due to information failures and search costs. Combining a tax on domestic electricity use with energy-efficiency labels for appliances can enhance the environmental effectiveness of *both* instruments. On the one hand, the label could increase the effect of the tax – by increasing the relevant price elasticities. On the other hand, the introduction of a tax would often increase the attention households, firms and others pay to a related label.

The importance of information failures and the need to address them are, however, not the same for all environmental issues. They are likely to be most important:

● when the up-front investment costs of an efficient product are relatively high;

● in markets that are characterised by low competition;

- in markets where split-incentives are prevailing, like between landlords and tenants in parts of the housing markets;
- when significant environmental damage could be caused if no intervention is made.

The mutual reinforcements of the instruments is likely to be most important if there are significant *private benefits* associated with a change in behaviour, through the purchase of (favourably) labelled products. The existence of any private benefits depends on the environmental issue addressed. For instance, users do get a direct private benefit from using energy-efficiency products, through lower operating costs. They can also save operating costs by using water-efficient products – when water-use is metered – but the magnitude of these benefits will generally be lower than those associated with energy use.

On the other hand, there are hardly any private benefits associated with buying certain products just because they are part of a recycling scheme – unless, for example, the size of a tax or charge levied on that product depends on the participation in such a scheme. Hence, recycling-related logos are less likely to increase the environmental effectiveness of other instruments in a given mix.

In addition to differences between environmental issues, the impacts of such instrument combinations will also vary between different products. For instance, the increase in the relevant price elasticities due to the introduction of a label is likely to be more significant for some electrical appliances (*e.g.* a television or a refrigerator) than for a motor vehicle, for which the energy consumption in any case is likely to be a product characteristic taken into account by households in their decision-making, at least to a certain extent.

It is also important to note that applying *several different labels* to target a given environmental objective (*e.g.* energy efficiency of appliances) can create confusion in the message sent to the consumer. In addition, it may entail high administrative costs.

While there are many studies that evaluate the efficiency and effectiveness of various labelling or certification programmes *in isolation*,[18] much less is known about the (marginal) impacts *in practice* of *combining* such schemes with a tax or charge. However, Jänicke *et al.* (1998) studied the impacts of the combination of *i.a.* rapid increases in electricity taxes over several years and a labelling scheme indicating the fuel efficiency of refrigerators in Denmark in the 1990s, and found that the two instruments did in fact under-pin each other. The impacts of both instruments were further enhanced by *e.g.* special training provided to about 20% of all sales staff connected with retail sales of "white goods".

10.5. Taxes used in combination with negotiated agreements[19]

10.5.1. *Introduction*

When taxes are used to address an environmental problem, firms' compliance costs are equal to abatement costs *plus* tax payments for residual emissions. A number of countries combine certain taxes or charges with voluntary schemes, where for instance some sectors are completely exempted from a tax – or pay lower tax rates than other sectors – on the condition that they "voluntarily" undertake certain abatement measures. Such arrangements are often introduced due to a fear that the international competitiveness position of the firms concerned would be compromised if they had to pay the full tax rate. If this position was significantly weakened, plant closures could result, with subsequent transition costs related to capital losses and increases in unemployment, *sometimes* in regions with limited employment opportunities.

One should consider what would *realistically* be the alternative policy when discussing the impacts of combining an environmentally related tax or charge with, *e.g.*, tax exemptions in return for negotiated agreements with some firms or sectors. If the alternative policy is a flat tax rate for all relevant polluters, at the same ("high") level as used for some sectors in the combined policy, the introduction of a voluntary option for some polluters would likely represent a weakening of the environmental target and/or a lower degree of achievement of a given target. Even if a negotiated agreement would oblige the polluters to abate emissions – and leave them increased financial resources to invest in pollution abatement, through the forgone tax revenue – it is not given that this would outweigh the emission reductions that "ordinary" price responses, and possible plant closures, under a "full tax regime" would have brought about.

Impacts on technology development could also be important: adding the voluntary option could give the affected firms more financial resources to undertake research and development, but their incentives to *actually achieve* technology improvements – and their profits from doing so – could be severely reduced. When the "shadow price" on emissions approaches zero, the firms have low incentives to find ways to reduce them. Over the longer term, this could have important environmental repercussions.

If the realistic alternative to a voluntary approach is a much lower tax rate for the firms included in the voluntary approach than for other firms, the significance of the points above would be reduced accordingly.

In both cases – but to a varying degree – replacing a tax by a voluntary approach will induce a revenue-loss for the government. As discussed by *e.g.* Fullerton and Metcalf (2001) and Goulder, Parry and Burtraw (1997), this revenue-loss can represent a significant efficiency cost. The scarcity rents created by the environmental policy are left with the private companies. Public authorities could, for instance, have used the revenues foregone to lower distorting taxes on labour, thus stimulating employment.

Various types of administrative costs could increase with the introduction of a voluntary scheme in addition to a tax. Most environmentally related taxes are relatively simple to administer, with *e.g.* the tax-bases being measured and revenues being collected at a limited number of oil refineries for most taxes on mineral oils. Introducing a conditional tax reduction can significantly increase the administrative burden, both for public authorities and for the firms involved.

The following sub-sections discuss a few concrete examples of combinations of taxes or fees with voluntary approaches.

10.5.2. The CO_2 tax and the Energy Efficiency Agreements in Denmark

In Denmark, a policy package implemented in 1996 combined the introduction of SO_2 and CO_2 taxes with an agreement scheme on improved energy efficiency in industry, and subsidies for *e.g.* energy efficiency counselling and investment.[20] All revenues from the taxes were recycled back to industry in the form of reductions in the taxation of labour and through subsidies for energy efficiency measures. Firms that entered into an agreement with the Danish Energy Agency got a rebate on their CO_2 tax. While all firms with *heavy processes* had the right to enter into an agreement, firms with *light processes* only had the right to sign an agreement – and get a tax rebate – if the tax payment on their energy consumption amounted to at least 3% of value added. In addition, the effective tax had to exceed a certain minimum value.

The agreements could be either individual or collective, covering several firms within a sub-sector with similar production processes. The basis of individual agreements *used to be* an energy audit, usually carried out by a certified consultant. As from 2000, simpler energy surveys have replaced the energy audits.

The collective agreements were not based on energy audits. Instead, an analysis of energy consumption and production processes in the sector was made to identify general potentials for improving energy efficiency in the relevant firms. The results of the analysis were reported to the Danish Energy Agency and used to formulate an action programme. In addition to investment projects, special investigations and energy management measures, the action programme for the sub-sector could include inter-firm projects, such as development projects, which were of interest to all firms. Each firm covered by the agreement had to sign and was committed implement all identified energy saving projects with a payback-period of less than 4 (heavy process) or 6 (light process) years. Firms also had to introduce improved energy management systems.

The use of a payback-period criterion implies that firms with many profitable investments would have to realise relatively large energy savings, while firms with no profitable projects were not burdened with investment projects and special investigations. This contributes to an efficient allocation of energy savings between firms.

However, there were important differences in the criteria used for different firms. Firms with light processes used to be required to undertake projects with longer payback periods than firms with heavy processes. In addition, different price assumptions were used when calculating the payback-periods: For firms with heavy processes, a (hypothetical) tax of EUR 3.3 per tonne CO_2 was added to the pre-tax energy price of the firm, while for firms with light processes, a (hypothetical) tax of EUR 12 per tonne CO_2 was added. The lower the tax being applied in the analysis, the lower is the likelihood that a given project would pass the test. Hence, some relatively low-cost energy-saving projects in firms with heavy processes could be left unrealised, which would tend to increase the overall abatement costs for society as a whole.

It is emphasised that the agreements provided limited *additional* tax reductions to the participating companies compared to the very large tax reductions granted to any industrial firm that employ light or heavy processes, see Table 10.2. However, a reduction in the tax rate for, *e.g.*, firms with heavy processes from EUR 3.3 to EUR 0.4 per tonne CO_2 in 2000 was, of course, in itself substantive.

Table 10.2. **Levels of CO_2 – and energy taxes in Denmark**

1996-2000, € per tonne CO_2

	1996	1997	1998	1999	2000
Space heating[1]	26.7	53.3	80	80	80
Light processes					
Without agreement	6.7	8.0	9.3	10.7	12.0
With agreement	6.7	6.7	6.7	7.7	9.1
Heavy processes					
Without agreement	0.7	1.3	2.0	2.7	3.3
With agreement	0.4	0.4	0.4	0.4	0.4

1. The numbers represent the total energy and CO_2 tax rate for space heating. The CO_2 tax rate alone was EUR 13.4 each of the years 1996-2000.
Source: OECD (2003a).

The energy efficiency agreements had two opposing effects on energy use in the respective companies. On the one hand, the companies had to carry out certain activities, like realising proposed energy-saving projects from the energy audits described above, and to increase energy management activities. The effect of these activities was estimated by Bjørner and Jensen (2002) to be a 9% *reduction* in energy use in the companies concerned. On the other hand, companies with an agreement obtained a tax reduction, which was estimated to *increase* their energy use by some 1-5%. Hence it appears that the agreement scheme resulted in a reduction in energy use overall. In other words, the agreement companies would, according to the findings of Bjørner and Jensen, have used more energy if they had not been offered the agreement, but just had paid the "normal" tax.[21]

10.5.3. The Climate Change Levy and the Climate Change Agreements in the United Kingdom

Before the introduction of the Climate Change Levy in United Kingdom in 2001, energy-intensive sectors were given the option to obtain an 80% reduction in the tax rate if they entered into Climate Change Agreements on improving energy efficiency or reducing carbon emissions.

Agreements were made with 44 sector associations, covering more than 5 000 separate operators and 10 000 facilities.[22] They were negotiated with the relevant sector trade associations on behalf of the companies within the sectors concerned. Facilities identified in the agreements were eligible for the 80% tax discount until 31 March 2003. Eligibility for discount from 1 April 2003 depended on whether the first targets set in the agreements were met. The agreements span the period up to 2010, with "review points" in 2004 and 2008.

The agreements set targets both for sectors and for each separate facility. Some sectors use a common percentage reduction target for all facilities concerned, while other sectors have internally negotiated other ways of sharing the burden. If a sector as a whole fulfils its target, each facility in that sector is deemed to be in compliance. If a sector fails to meet its overall target, those facilities that have not met their own targets loose the 80% tax discount for the next 3 years.

The fact that it is enough for the sector to meet the overall target for all the facilities to maintain their discount could – in isolation – stimulate "free-riding", where under-performing facilities would try to benefit from abatement efforts at other plants. However, facilities that do better than required have the possibility to sell the surplus reduction into the UK CO_2 Emission Trading Scheme. Hence, in practise, each facility must make sure that they meet their own target.

There has been considerable debate as regards the impacts of the agreements. de Muizon and Glachant (2004) provided a theoretical discussion of the combination of the Climate Change Levy, the Climate Change Levy Agreements and the UK emissions trading system. They concluded that the performance of this instrument mix would not have been affected by an absence of the agreements.

Cambridge Econometrics (2005) wrote *inter alia*:

"A combination of technological change and relative decline in UK energy-intensive subsectors of manufacturing (ie bulk chemicals as opposed to specialty chemicals), implies that the energy (and therefore carbon) saving and energy-efficiency targets would have been met without the CCAs. This result is uncertain because the historical technical and structural-change trends may not continue as in the past, ... Moreover, the CCA targets are set in terms of improvements in energy efficiency, whereas the model projections have used energy intensity which means that the comparison is distorted with any structural change within the sectors. Only for one sector (other industry in 2008) did we find that the CCA target would have been missed had no CCL ever existed. We also found that the price effect of the reduced-rate CCL was sufficient, on its own, for the target to be met ...”

To the extent it is correct that the targets set under the Climate Change Agreements would have been met in any case, the environmental effectiveness of adding a voluntary approach to a tax – even if reduced rates were applied for some sectors – would be very modest, largely limited to any “awareness-raising” effects and similar they might have.

10.5.4. The Intention Agreement on SO_2 emission reductions in Norway

A tax on the sulphur content of fuels has been in place for many years in Norway, covering at present about 27% of all SO_2 emissions in the country – with a tax rate of approximately EUR 2 per kg SO_2. From the outset, emissions from refineries, from the use of coal and coke, the use of mineral oils in the petroleum extraction activity on the continental shelf, and from supply-ships of this activity, were completely exempted from the tax. In 1999 these emission sources were included in the tax, with a reduced tax rate of about EUR 0.35 per kg SO_2.

However, from 1.1.2002 emissions from refineries and from the use of coal and coke (largely in industrial processes) were once again completely exempted from the tax. In return, the Federation of Norwegian Process Industries had signed an “Intention Agreement” with the Ministry of Environment, committing the member companies to reduce SO_2 emissions by 5 000 tonnes by 2010, and to prepare a plan on how emissions could be reduced in a cost-effective way by a further 2 000 tonnes. Together this would equal the total emission reductions Norwegian authorities expected the country had to make to fulfil its obligations under the Gothenburg protocol of the UN ECE convention on Long-range Transboundary Air Pollution, capping total Norwegian SO_2 emissions at 22 000 tonnes from 2010.

Studies undertaken by the Norwegian Pollution Control Authority indicate that the most cost-effective measures to reduce SO_2 emissions in Norway can be found in the process industry. In 1999, firms taking part in the agreement emitted more than 16 000 tonnes of SO_2, compared to total Norwegian emissions of some 29 000 tonnes.

Norwegian SO_2 emissions have already decreased considerably since 1999, inter alia due to plant closures and temporary production reductions in the process industries.

The Federation of Norwegian Process Industries have stated that the tax rate of about EUR 0.35 per kg SO_2 was not environmentally effective, as it was cheaper for firms to pay the tax than to install cleaning equipment that would be required to reduce emissions. This was underpinned by the findings of SFT (2001), where all potential abatement measures (with one exception) were found to have a marginal cost of EUR 0.45 or more per kg SO_2 abated. The other cheapest measures to reach a 7 000 tonnes emission reduction in total were found to have marginal costs of between EUR 1 and EUR 1.5 per kg SO_2 abated.

The Federation of Norwegian Process Industries further stated that if the tax rate had been set so high that it would be profitable for the firms in question to install the cleaning equipment, the firms would not survive economically.[23]

The "Intention Agreement" is *not legally binding* for the two parties. Until the measures covered by the agreement have been implemented, by 2010 the latest, the ordinary environmental emission permit system will be the main policy instrument addressing the emissions from the sources concerned.

The process industry has – based on a *legally binding* "implementation agreement" involving all the firms that used to pay the lower tax rate – set up an "environmental fund", organised as a self-owned foundation, and financed by fee payments similar to the previous tax. An "action plan" for how the Intention Agreement is to be fulfilled was developed in 2003. It is estimated that the fund will raise about EUR 30 million in revenues through the fees paid by the member companies. The resources of the fund will be used to – fully or partially – finance development, implementation and operation of abatement measures and other measures suitable in the pursuit of the targets of the implementation agreement, including support to closure of activities that leads to lasting emission reductions. Measures are to be implemented where they will contribute the most to reduce emissions, until the targets of the Intention Agreement have been reached. Consideration will also be given to where emission reductions will contribute most to improve local air quality. In general, measures will be supported based on applications from the participating firms. If not enough applications should be made to reach the targets of the Intention Agreement, a site might be *instructed* to undertake a measure financed by the fund.

It seems that the most realistic alternative policy to this combination of a negotiated agreement and the sulphur tax being applied to other sectors in Norway would have been a reduced rate in the sulphur tax for industry. And as it seems less costly for the firms to pay the tax rate that was applied between 1999 and 2002 than to abate emissions, the new instrument mix – which also included industry-financed and industry-managed subsidies for abatement measures – might lead to lower emissions from industry.[24] Similar emission reductions *could* have been obtained by increasing the tax rate sufficiently to make it cheaper for firms to abate than to pay the tax, but that might have entailed significant social costs – at least in the short to medium term – to the extent threats about plant closures are correct.

It is also important to emphasise that mechanisms to promote economic efficiency under the agreement have been put in place in this case. Instead of *e.g.* every firm reducing their emissions by equal percentage amounts, the firms involved agreed to "pool" resources, and to undertake the emission reductions where they can be obtained at lowest cost. The decisions on where to make the emission reductions will be taken by representatives of the firms involved – who should have better knowledge of the actual costs of various abatement options than public authorities in many cases will have.[25]

10.5.5. Conclusions on combinations of taxes and voluntary approaches

Both the energy efficiency agreements in Denmark, the Climate Change Agreements in the United Kingdom and the "Intention Agreement" on SO_2 emission reductions in Norway have – to a significant degree – been motivated by a wish to prevent close-down of industrial plants that could have taken place if "full" tax rates had been applied. It seems unlikely that the agreements provide environmental benefits beyond what "full" tax rates would have done, but, in the case of Norway, it seems that the previous reduced rates applied to certain industrial sectors were too low to have any significant environmental impact – at least in the short term.

By combining taxes and a voluntary approach in these cases, policy makers have tried to avoid having to make trade-offs between the environmental, economic and social dimensions of sustainable development. It remains to be seen whether such trade-offs can be avoided in the longer term, as – for example – more ambitious climate policies are being put in place.

As stated above, under a tax regime, firms' compliance costs are equal to abatement costs plus tax payments for residual emissions. When firms can avoid paying for any residual emissions by taking part in a voluntary scheme, impacts of the policy on the production costs of these firms will be limited. A wish to limit such cost impacts – especially for firms facing stiff international competition – is exactly one of the reasons for which the tax relief is given. However, to the extent the firms could have shifted any part of the cost increases on to their customers – through increases in the prices of their products – applying a voluntary approach looses out on any direct impacts on the demand for products that cause pollution in their production. In many cases, such demand changes could provide an important part of the environmental benefits from using economic instruments.

10.6. Taxes used in combination with subsidies

It can be useful to distinguish between two main cases when discussing the impacts of combining a tax and a subsidy scheme:

1. one case is when the revenues from an environmentally related tax are earmarked to provide subsidies for certain environmental purposes;

2. another case is when a tax and a subsidy scheme are combined, without any explicit link between the revenues raised by the tax and the financing of the subsidy scheme.

10.6.1. Earmarking of tax revenues for environmental subsidies

As mentioned in Section 2.1, the revenues of about 1/3 of the 375 or so environmentally related taxes in OECD countries are earmarked for a particular purpose. About 75 of the earmarked taxes are levied on energy products, 15 are levied on motor vehicles while 20 are waste-related taxes. Whereas the earmarked on motor fuels or motor vehicles tend to be allocated to the construction or maintenance of roads (sometimes including for environmentally benign purposes like the building of noise-protection walls, development of bicycle lanes, improvements in public transport, etc.) earmarked waste-related taxes are normally used more generally for environmental purposes, in particular for the operation of waste collection or recycling systems, for the clean-up of contaminated sites, etc. Also a number of charges are earmarked for various environmental purposes, including for subsidies for specific issues.

The focus in this chapter is on cases where revenues are earmarked for environmental purposes. Some of the findings below are, however, relevant also when the earmarking is for other purposes.

There might come some positive environmental impacts from the earmarking of tax revenues for environmental purposes. For example, Andersen (1999) concluded in his comparative study on the clean water policies in four OECD countries that one factor behind the success of the Dutch system of water-related charges, compared to the Danish and Belgium systems, was the earmarking of revenue. The process by which revenues were earmarked in the Netherlands improved co-operation among polluters and between specialised water institutions and polluters and regulators. This may have resulted in lower transaction costs in implementing and investing in effective pollution reduction. Andersen argues that the "availability of information and advice, and the opportunities for financial assistance on the basis of the proceeds of the levies, smoothed the transition" to a reduced pollution outcome.

However, earmarking revenues fixes the use of tax revenue in advance, which may create an obstacle for a later re-evaluation and modification of the tax and spending programmes. Programmes may last longer than is optimal because of bureaucratic and other vested interests' obstruction to reform, even when policy priorities have changed. Therefore, the economic and environmental rationale of such measures should be evaluated regularly to *avoid inefficient spending* that would otherwise not be financed from general tax revenues.

Earmarking revenues may in some cases improve the political acceptability of taxes because of the dedicated nature of expenditures, a proportion of which is returned to taxpayers in the form of subsidies or public investments. If such impacts are deemed necessary for the implementation of a new (otherwise well-motivated) tax, a compromise solution could be to simultaneously with the proposal of the tax to propose increased expenditure for a relevant, "popular", purpose. This can be done without creating a legally binding link between the size of the revenues and the size of the expenditures in future years.

10.6.2. *Other combinations of taxes and subsidies*

Disregarding now the issue of earmarking, environmentally related taxes can still be used in combination with various types of subsidies. The focus in this chapter is on combinations of instruments for environmental policy, so the most relevant subsidy category to consider here is "environmentally motivated subsidies". More perverse cases do, however, exist – when the taxes are combined with "environmentally harmful subsidies", see Box 10.1. An example of such is the use of taxes on fossil fuels in combination with subsidies for coal mining – or for the extraction of other fossil fuels for that matter. Another example is taxes on pesticides and/or fertilisers in combination with subsidies that promote intensive farming. A strong candidate for policy reform is to scale back or otherwise modify existing environmentally harmful subsidies.[26]

More to the point here are cases when one environmentally related tax is combined with a subsidy to *promote* one or more environmental impact(s). For example, taxes on fossil fuels are in several countries combined with subsidies to produce bio-fuels or other renewable energy sources – as a means to address *e.g.* climate change (and "fuel security").[27] Taxes and charges in the waste area are often combined with subsidies for the establishment and/or operation separate waste collection schemes and/or for various (other) recycling activities.

Whereas the overall impact might be larger than if only one instrument had been applied, it is difficult to see any strong *synergy effects* between the instruments in these cases. That is, the subsidies do not make the tax work any better *per se*, or *vice versa*. The same overall impact could in many cases have been obtained by increasing somewhat the tax rates applied to the relevant environmental harm(s) – which could be beneficial from a budgetary point of view.

Split incentives between landlords and tenants can be an obstacle to the environmental effectiveness of a tax on domestic energy use. Better isolation of the

---

Box 10.1. **Defining subsidies**

The concept of subsidy is not straightforward. While the term "subsidy" is used in this paper, it is as common to use the terms transfers, payments, support, assistance or protection associated with governmental policies in OECD work. Sometimes these terms are used interchangeably, but often they are associated with different methods of measurement and thus different economic indicators. Subsidies have been defined to "comprise all measures that keep prices for consumers below market level or keep prices for producers above market level or that reduce costs for consumers and producers by giving direct or indirect support" [see, for example, de Moor and Calamai (1997)]. This definition is consistent with the OECD approach of defining environmentally harmful subsidies and tax concessions to include "all kinds of financial support and regulations that are put in place to enhance the competitiveness of certain products, processes or regions, and that, together with the prevailing taxation jurisdiction, (unintentionally) discriminate against sound environmental practices" (OECD, 1998). It is not necessary to make a distinction between subsidies and tax expenditures as the latter can be regarded as implicit subsidies.

Subsidies take different forms: budgetary payments or support involving tax expenditures (various tax provisions that reduce the tax burden of particular groups, producers or products), market price support, subsidised input prices, preferential interest rates. This is why the more generic terminology of "*support measures*" is often used. There is, however, no international consensus: different definitions prevail for specific purposes, fields (*e.g.* agriculture or transport) or contexts (*e.g.* international trade).

There has been much controversy over whether the non-internalisation of external costs should be construed as a subsidy, the argument being that, as external cost are not internalised, the environment is used "freely" by the users: in a sense, a public good is freely supplied to users. Those who object to such an expanded definition observe that the notion of a subsidy has traditionally connoted an explicit government intervention, not an implicit lack of intervention. As well, for these and more practical purposes, namely the difficulty of quantifying external costs, non-internalisation is generally not regarded as a subsidy, except for the transport sector where this definition is currently used (Nash *et al.* 2002).

Source: Barde and Braathen (forthcoming).

---

dwelling could be a cost-effective measure to reduce the energy use, but the landlord could be reluctant to undertake such investments, as the benefits in the form of energy savings would mostly accrue to the tenant.[28] On the other hand, the tenants could be reluctant to –and perhaps legally restrained from – undertaking such investments on their own, as the benefit of a higher value of the building would accrue to the landlord. Combining a tax on energy use with targeted subsidies for such investments can be a way to *address split incentives*.

In general, when trying to reflect differences in environmental impacts in the relative prices through the use of various subsidies there is a danger of coming in conflict with the "Polluter-Pays Principle". According to the OECD (1972 and 1974), the Polluter-Pays Principle means that the polluter should bear the "costs of pollution prevention and control measures", the latter being "measures decided by public authorities to ensure that the environment is in an acceptable state". In other words, the polluter should bear the cost of steps that he is legally bound to take to protect the environment, such as measures to reduce the polluting emissions at source and measures to avoid pollution by collective treatment of effluents from a polluting installation and other sources of pollution.

10.7. General findings on impacts of combining taxes with other instruments

The discussion above has indicated that in a number of cases there can be environmental and/or economic benefits from combining a tax with other types of policy instruments. In practice, environmentally related taxes are seldom used in complete isolation – in a large number of cases one or more regulatory instruments will, *for example*, be applied.

The mere existence of instrument mixes is, however, obviously not a "proof" of their environmental effectiveness and economic efficiency. OECD's work on instrument mixes used for environmental policy has indicated a number of requirements for developing effective and efficient instrument mixes, and these are briefly highlighted below.

10.7.1. A good understanding of the environmental issue to be addressed

A rather obvious first requirement for applying an environmentally efficient and economically effective instrument mix is to have a good understanding of the environmental issue to be addressed. In practice, many environmental issues can be more complex than perhaps first thought, as they often have a number of *relevant*, and often *correlated*, "aspects" or "characteristics" – and many of the instruments that are applied contain a large number of separate "*rules*" or "*mechanisms*". A simple counting of the number of targets and instruments is, hence, difficult – and perhaps sometimes of limited relevance.

For example, regarding the issue of pesticides use, while some of the environmental problems are relatively directly related to the *total amounts* of pesticides applied, other problems are more dependent on *which* pesticides are used, on *how* the pesticides are applied, on *where* they are applied or on *when* they are applied. In addition, the environmental outcomes will in many cases depend on the weather conditions "on average" in a given year, the weather conditions when the pesticides are being applied, the type of soil to which the pesticides are applied, local hydrological conditions, etc. A tax (or a tradable permit system) can relatively well affect the total amount of pesticides and the choice between different types of pesticides, but is less suited to address the other problems listed. Hence, other instruments could in any case be needed.[29]

On the other hand, in some cases it can seem that more environmental targets than necessary have been defined. This is perhaps in particular the case in the waste area, where – for example – specific recycling targets for a large number of products or waste streams have been established in many OECD countries, frequently without a clear documentation that the selected waste streams represent a larger threat to the environment than other, related, waste streams.[30]

10.7.2. A good understanding of the links with other policy areas

To ensure the coherence of the instrument mix applied in a given environmental area, different levels of policy co-ordination will be needed. In addition to co-ordinating different environmental policies, co-ordination with other related policies is needed.

For instance, in the area of residential energy efficiency, co-ordination with building regulations, housing policy and tax policy (*e.g.* property-based taxation) is needed to ensure instrument mixes' coherence. This may in particular be necessary to address market imperfections such as "landlord/tenant failures" and to deal with social concerns.

Also as regards household waste is co-ordination with other policies required. One paradox in this connexion is the fact that emission regulations for several pollutants facing waste incinerators tend to be stricter than the emission regulations facing other emitters – *cf.* for example the fact that only waste incinerators face the new tax on emissions to air in Norway.

10.7.3. A good understanding of the interactions between different instruments

A first observation that can be made is that any instrument mix risks being in-effective if "correct" price signals are not transmitted to the relevant decision-makers. The household waste area can again serve as an example: Despite a broad spectre of long-lasting policies addressing (*inter alia*) total amounts of household waste in the Netherlands – and more recent measures in the same vein in United Kingdom – total household waste amounts are as high in the Netherlands as they are in United Kingdom, and they have increased over time in both countries. *In part* this is likely to be because only some 20% of the households in the Netherlands – and *none* of the households in the United Kingdom – face a waste collection charge that varies with the amount of waste they generate. Hence, the landfill taxes applied in both countries do not give households a clear incentive to generate less waste.[31]

Various instruments can interact with environmentally related taxes in a number of ways. *For example*:

● As discussed in Section 10.4, a labelling system can help increase the effectiveness of a tax, by *providing better information* to the users on relevant characteristics of different product the tax applies to. The price elasticities of concern can hence increase.

● As mentioned in Section 10.6, combining a tax on energy use with targeted subsidies for better isolation of buildings can be a way to *address split incentives*.

● As addressed in Section 10.5, the combination of a tax and a voluntary approach can *increase the "political acceptability"* of the former – by limiting any negative impacts on sectoral competitiveness – at the cost of reduced environmental effectiveness or increased economic burdens placed on other economic actors.

● As shown in Section 10.3, combining a tax and a tradable permits system can help *limit compliance cost uncertainty* – compared to the application of a trading system in isolation.

● On the other hand, such a combination would *increase the uncertainty related to the environmental effectiveness*.

● As pointed out in Section 10.2, there is also a danger that a regulatory instrument applied next to an environmentally related tax could *unnecessarily restrain the flexibility* for polluters to find cost-effective abatement options offered by a tax.

Notes

1. For further information, see OECD (forthcoming).

2. It is also known from the "theory of the second best" that in general one needs as many policy instruments as there are market failures in order to achieve an "optimal" outcome.

3. *E.g.* the Czech Republic, Denmark, France, Korea, Norway, Sweden and Switzerland.

4. The exact rationale for applying two tax instruments to address sulphur emissions – introducing one "shadow price" for SO_2 emissions in general, and a still higher "shadow price" for the SO_2 emissions that stem from the use of petrol and diesel – is not quite clear. The harm done by SO_2 emissions from motor vehicles is generally the same as the harm done by SO_2 emissions originating from other sources. A possible motivation is, however, to give petroleum refineries an incentive to produce so clean fuels that car manufacturers can develop motors that limit the emissions of a number of *other pollutants* as well.

5. See CEU (1999).

6. This section is largely based on OECD (2003b).

7. The effects of uncertainty with respect to the marginal benefits of abatement do not have analogous effects, since firms will respond equivalently in the two cases.

8. See Jacoby and Ellerman (2004) for a discussion of a "safety valve" in the context of climate policy.

9. In the final Directive, the penalty in the first period is set to EUR 40 per tonne, see EU (2003a).

10. Conversely, in the UK's landfill permit scheme for biodegradable municipal waste, the government considered the possibility of introducing a cap through a penalty/tax, but felt that this was unnecessary since there was little uncertainty concerning the costs of waste diversion (UK DETR 2001).

11. A possible reason for why the tax is not used as a "safety valve" in this case is that the Landfill Allowance Trading System was recently added to the Landfill Tax with the explicit purpose of securing compliance with the limitations on the amounts of waste going to landfilling, according to the EU's Landfill Directive. Introducing a 'safety valve' would increase the uncertainty related to the fulfilment of the environmental objective.

12. See *www.epa.gov/airmarkets* for data on the efficacy of monitoring in the program. Separate sanctions can be introduced for tampering with the monitoring equipment.

13. See DETR (2001) for a discussion of the principles used to determine the level of the penalty under the Landfill Allowance Permit Scheme in United Kingdom.

14. In this vein, it is important to note that this rent was already being received under pre-existing regulatory systems through issuance of permits (see Fullerton and Metcalf 1997 for a discussion). However, it was not convertible as an asset. This is not to say that it was less valuable. Indeed, in some cases, it may have been more valuable than grandfathered permits. For instance, the new-source bias in EPA pollution regulations meant that existing permitted firms received the rents associated with high barriers to entry for new firms.

15. It is worth noting that EU member states have mostly maintained any energy taxes they apply to firms that form part of the EU-wide greenhouse gas emission trading system that was introduced in 2005. Whereas much of the industry objected to the introduction of this trading scheme, the firms affected were in fact given a significant economic rent for free.

16. See Johnstone (1999) for a further discussion.

17. To the extent increased price responsiveness means that a given environmental target can be achieved through a smaller change in relative prices, the economic efficiency of the instrument mix would also be enhanced.

18. See, for example, Gillingham, Newell and Palmer (2004) for a recent literature review concerning non-transportation energy efficiency policies.

19. This section draws largely on OECD (2003c) and Braathen (2005).

20. See OECD (2003a) for more details. The study focused on how the Energy Efficiency Agreements were designed up to 2000.

21. The estimated increase in energy use due the tax reduction was 5% in 1993 and 1995, but only 1% in 1997, when the tax reduction offered to participating companies was much smaller. In later years, the tax reduction has once again increased, which would tend to add to the associated increases in energy use. It is underlined that the calculations of Bjørner and Jensen (2002) only take into account the specific tax concessions granted to firms participating in energy efficiency agreements. The impacts of the – much bigger – tax reduction granted to all firms that employ light or heavy industrial processes were not estimated.

22. Where an energy-intensive installation uses less than 90% of the energy within a site, the facility which is covered by a Climate Change Agreement must be sub-metered, so that the energy used by the facility is known accurately. According to DEFRA (2001), costs may be in the region of GBP 1 000 to GBP 5 000 per meter, possibly more where the energy supply arrangements are particularly complicated. In sectors such as supermarkets and aerospace, a programme of installing sub-meters was agreed. Both the supermarkets and master bakers sectors have *over a thousand sites* with activities which are eligible to be covered by an agreement and where additional metering is required if they are to be included in an agreement. This is one reflection of the *considerable* administrative costs negotiated agreements can entail.

23. Several emission-intensive plants have nevertheless been closed since 1999 – for reasons not related to the Intention Agreement. Rather than retrofitting plants that anyway have a modest or low profitability, it can be a low-cost option for society as a whole to reach environmental targets by closing down some of these plants.
According to the Federation of Norwegian Process Industries, emissions from the plants covered by the agreement increased 1 700 tonnes, or 13.7%, between 2003 and 2004 – due to an improved business cycle situation.

24. The environmental target for the country as a whole is in this case given by the obligations under the Gothenburg protocol. The issue at stake is to what extent the emission reductions should be undertaken in the process industry – where abatement costs are the lowest – but where there is also a fear for the international competitiveness position of the firms concerned. It should, however, be emphasised that the protection of the sectoral competitiveness of the process industry comes at the price of increased costs – and lower competitiveness – for the remainder of the Norwegian economy.

25. In this particular case, the Norwegian Pollution Control Authority had made detailed studies of the costs of various abatement options, *cf.* SFT (2001). While the findings of these studies very well could be correct, there is in general an asymmetry in the information available to the firms involved and to public authorities.

26. For more in-depth discussion of environmentally harmful subsidies, see OECD (2003e, 2005c and 2006b).

27. It should, by the way, be kept in mind that the production of bio-fuels in itself requires a significant amount energy use – with a number of other negative environmental impacts induced as well, *e.g.* related to the use of pesticides and fertilisers, etc.

28. This is in particular the case when the rents a regulated.

29. As always, one should consider whether the benefits of applying additional instruments would exceed the related costs.

30. The waste area is also illustrating a frequent lack of a proper, life-cycle based, economic valuation of the costs and the benefits related to the policy targets set, and to the instrument mixes being applied.

31. It can seem that too much attention is given to *total waste amounts* in many member countries. It would be more appropriate to focus more explicitly on the *different negative environmental impacts caused by different waste streams*.

TECHNICAL ANNEX

Estimation of Price Elasticities

(Terms in *italics* are generally explained somewhere in the text.)

Price elasticity can be defined as the behavioural response to a change in prices. For example, one can estimate the price elasticity of residential energy use in analysing how household demand varies with respect to variations in the prices of domestic fuel and electricity. Price elasticities have been estimated using different types of data (*e.g.* time-series, cross-sections and panel data) and different estimation approaches. As different approaches provide different estimates of the variables of interest, it is important to know which estimation technique is the most appropriate according to the type of datasets and considered models.

A.1. The type of data

Three types of data can be used to estimate price elasticities: cross-sectional data, time-series data or panel data. These data types can be defined as follows (Greene, 1997):

- A *cross-section* is a sample of a number of observational units all drawn at the same point in time.
- A *time series* is a set of observations drawn on the same observational unit at a number of (usually evenly spaced) points in time.
- *Panel data* are a combination of both cross-sectional and time series data and consist of a large number of observational units observed at more than one point in time, *i.e.* the same cross section is observed at several points in time.

An underlying assumption when using cross-sectional data is that the different firms and households have had time to adjust to the differences in prices they are facing. If important changes in the relative prices took place shortly before the cross-section was selected, this assumption would not be valid.

Although cross-sectional and time series data are widely used to estimate price elasticities, several reviews of the studies estimating elasticities of road traffic and fuel consumption (Graham and Glaiser, 2002; Goodwin *et al.*, 2004) suggest that most recent studies tend to use panel data instead of cross-sectional data. Two main reasons can explain why panel data are favoured over cross-sections. First, in panel data (and time series as well) time is explicitly entered into the model. As a consequence, panel data are less affected by any omitted variable bias and they provide a more flexible framework to analyse the behavioural differences across cross-sectional units.

Time series data does, however, have several drawbacks. First, as the full impact of price changes on the demand may take a long time to be complete, data over a long time-period are thus required. The gradual adjustment of the demand to price changes should usually be reflected in differences between estimates of short- and long-term elasticities. In addition, estimates can be very sensitive to the particular specification chosen and thus misspecification can seriously undermine the reliability of TS estimates. In addition, TS data are often subject to *autocorrelation*.

On the other hand, estimates from cross-sectional data are likely to be more robust to misspecification than time series and panel data estimates, as there is a greater variance in the latter than in the former. Cross-sections can thus provide better estimates of long-term effects than time series data. However, *heteroscedasticity* (and all associated problems) is most commonly expected in cross-sectional data.

Pesaran and Smith (1995) estimated long-term energy price elasticities using three types of data: cross-sectional, time series and panel data. Their results clearly highlight the fact that different approaches provide different values. In their example, long-run cross-section estimates were in general the highest; time series and weighted averaged time series (*i.e.* estimates averaged over the groups) estimates provide the lowest estimates; and, panel data estimates lie between cross-sectional and time series estimates. However, panel data impose the restriction that the coefficients are all the same. This assumption induces *serial correlation* and will produce inconsistent estimates if lagged values of the dependent variable are included in the regression (Pesaran and Smith, 1995). They therefore recommend using time series data to estimate long-term energy price elasticities. This recommendation is echoed in Hanly *et al.* (2002). They highlight the particular relevance of using time series data within a dynamic framework to estimate short-term and long-term elasticities, especially in the transport area.

Standard econometric techniques are applied to cross-sectional data, time series data and panel data. Time series data and cross-sections are usually estimated by ordinary least squares (OLS) while panel data are estimated either via OLS (fixed effects model) or generalised least squares (GLS) (random effects model). A description of the different types of model and the different econometric approaches is provided in the next section.[1]

Differences between micro and macro data

Either micro or macro data can be used to estimate price elasticities. However, given their fundamental differences, the same econometric techniques cannot be used without some caution.

Micro data, such as the characteristics of a household or a firm, will always show a large variation in the variables. This heterogeneous nature of the data ensures that, in general, no major problems occur, such as *serial correlation* (*i.e.* correlation between the *dependent variable* and the error terms, which reflect everything we do not have information about) and estimation is relatively straightforward. However, macro data do not show such variation. Macro data always involve *non-stationary* and/or trending variables, such as income, consumption, price level, etc. Therefore, macro data often tend to move in the same direction because of some common trend. This phenomenon, which is generally associated with time series and panel data, has to be carefully taken into account when making the estimation.[2]

The usual way to deal with such a problem is to separately include a time trend in the model in order to remove the time effect in the other variables so their associated coefficient will reveal their true influence on the dependent variable. The use of such techniques, referred to as *cointegration techniques*, and associated models is recommended when using non-stationary time-series.[3]

Given the differences between micro and macro data, different econometric techniques are applied accordingly. Micro data can be estimated by all standard techniques: OLS, GLS and instrumental variables (IV) (if the dependent variable is continuous) or by maximum likelihood (ML) (if the dependent variable is discrete). As macro data may be subject to integrated variables or serial correlation, OLS should not be used, as they would provide biased and inconsistent estimates. It is then recommended to use either ML estimation or the GLS approach (Greene, 1997).

CGE-based estimates

Some price elasticities are estimated from CGE (computable general equilibrium) modelling, in particular in the trade area (see for example, Hertel *et al.*, 2004). However, several critics have highlighted the poor performance and econometric foundations of this approach (McKitrick, 1998). In addition, CGE models require (extremely) large data sets and CGE-based estimates are very sensitive to the choice of the functional form. Further work on CGE modelling is thus necessary to obtain reliable and robust estimates of price elasticities.

A.2. The type of regression model

Description of the different regression models

The classic cross-sections regression model is defined as the following:

$$y_i = \alpha - \beta'x_i - \varepsilon_i \quad \text{with } I = 1, 2, ..., N.$$

where α is a constant term, x_i represents the variables that characterise the ith observation and ε is the error term that catches everything that has not been observed.

The standard time series model is characterised by the following regression:

$$y_t = \beta_1 - \beta_2 x_t - \beta_3 y_{t-1} - \varepsilon_t \quad \text{with } t = 1, 2, ..., T.$$

This model is referred to as "autoregressive model", as it includes lag values of the dependent variable as explanatory variables. Another form of time series model, referred to as the "moving average" model, is characterised as follows:

$$y_t = \beta_1 - \beta_2 x_t - u_t - \lambda u_{t-1} \quad \text{with } t = 1, 2, ..., T.$$

This occurs when the error term ε_t is not a *white noise*, and can be defined as the combination of two (or more) white noises.[4] A combination of these two models provides what is called an "autoregressive moving average" process, such as:

$$y_t = \beta_1 - \beta_2 x_t - \beta_3 y_{t-1} - u_t - \lambda u_{t-1} \quad \text{with } t = 1, 2, ..., T.$$

As mentioned before, panel data are a combination of cross-sectional and time series data. As such, the most common framework for the regression model is defined as follows:

$$y_{it} = \alpha_i - \beta'x_{it} - \varepsilon_{it} \quad \text{with } I = 1, 2, ..., N \text{ and } t = 1, 2, ..., T.$$

In this framework, α_i represent unobserved effects. One of the main assumptions of panel data models is that the coefficients β of the different explanatory variables x_{it} are identical for all individuals in the panel: $\beta_t = \beta$ for all i. Moreover, the individual constant terms α_i differ with respect to individuals. In this context, two cases can be distinguished: the case where the parameters α_i are deterministic constant terms (fixed effects model) and the case where α_i are assumed to be random variables (random effects model).

In the fixed effects model, the individual effects α_i are represented by constant terms, hence the name of *fixed effects model*. It is generally represented as follows:

$$y_i = \alpha_i - X_i\beta - \varepsilon_i$$

All parameters in this model are constant and in order to simplify the model, we assume that there is no time effect. This modelling allows for arbitrary correlation between the unobserved effect α_i and the explanatory variables x_{it}.[5]

The standard random effects model is defined as:

$$y_{it} = \mu - \beta'x_{it} - \varepsilon_{it} \text{ with } \varepsilon_{it} = \alpha_i - \upsilon_{it}$$

α_i are random unobserved effects accounting for individual heterogeneity. As they are assumed to be not correlated with x_{it}, they are integrated as a component of the error terms and they do not have to be estimated.

The *Hausman test* allows knowing what form the regression model should take. This test is based on the assumption that, under the null hypothesis of no correlation between a_i and x_{it}, the fixed effects and the random effects estimates should not differ systematically. In other words, the null hypothesis of the Hausman test means that the random effects estimator is correct. This test then allows for the comparison of the two estimates. If they differ, it would mean that the null hypothesis is rejected and that correlation exists and is significant. This result would imply the necessary adoption of the fixed effects formulation.[6]

Non-stationary time series

As mentioned previously, most economic variables, such as GNP and consumption, exhibit strong trends and are not weakly stationary. A time-series is said to be weakly non-stationary when the mean and the variance of the data are not constant over time. They are often referred to as stochastic trending data.

In many cases, these series can become stationary by simple differencing. If the series y_t becomes stationary after being differenced one time, y_t is said to be integrated of order one, denoted I(1). A series denoted I(d) means that it requires to be differenced d times before becoming stationary. I(0) means that the series is stationary.

The standard approach for taking into account integration can be described as follows. The first stage consists in determining whether the series are non-stationary or not. This can be achieved in using the *unit root test* of stationarity (also known as the *Dickey-Fuller* test). For instance, consider the model:

$$y_i = \beta y_{t-i} - \mu_t$$

and estimate this model by OLS. If β takes a value greater or equal to 1, such that $y_i = y_{t-i} - \mu_t$, then the stochastic variable y_t has a unit root, and the time series is not weakly stationary. Alternatively, if β takes a value less than one, then the time series is stationary.

If the series are non stationary, the next stage consists in determining whether the variables are cointegrated or not. Consider the following regression model:

$$y_t = \alpha - \beta'x_i - \varepsilon_t$$

Generally, if two series y_t and x_t are both $I(1)$, there may be a β such that $\varepsilon_t = y_t - \alpha - \beta' x_i$. For example, it may happen if the two series are growing at roughly the same rate. Two series that satisfy this requirement are said to be cointegrated. In this context, we can distinguish between the determinants of the long-run relationship between y_t and x_t and those of the short-run relationship. If the time series are cointegrated, the long run elasticities are estimated directly from the cointegrating regression, defined as the first difference of the previous regression model, that is:

$$\Delta y_t = \alpha - \beta' \Delta x_t - \varepsilon_t \text{ with } \Delta y_t = y_t - y_{t-1} \text{ and } \Delta x_t = x_t - x_{t-1}.$$

An error correction mechanism model can be applied to estimate short-run elasticities. This modelling includes an error correction term which represents the long term equilibrium gap (*equilibrium error term*) (Greene, 1997).

Static vs. dynamic models

The elasticities can be estimated either within a *static* or a *dynamic* framework. However, the choice of the type of model will crucially depend on the nature of the available data.

As mentioned previously, cross-sections do not explicitly account for time. Therefore, it is quite obvious to use cross-sectional data within a static model. However, cross-sectional data can still provide long-term estimates of price elasticities. The problem is that this "long-term" timescale is rather indeterminate. On the contrary, time series and panel data, which explicitly account for time, are usually modelled within a dynamic framework but they can also be entered into a static model.

As time does not enter into static models, the variables are strictly exogenous and the coefficients vary randomly and independently with the variables. In this context, all econometric techniques can be applied, whatever the type of data used. Although the ordinary least squares (OLS) method is most commonly used, different approaches, such as generalised least squares (GLS) and maximum likelihood (ML) can provide consistent[7] and unbiased[8] estimates, with similar mean estimates of the coefficient (although the variances may differ). However, as static models only capture the response to a change in prices at a specific point in time, they may not fully capture the adjustment of demand to price changes. To this end, dynamic models are used to account for the potentially long-term adjustment between change in prices and changes in the household demand. The use of dynamic models is more attractive than the use of static models, as they allow for distinguishing between short-term and long-term effects.

However, the use of dynamic models is often associated with a set of problems, namely *endogeneity*,[9] *serial correlation* and *integration* issues. In this context, the different procedures of estimation will not give consistent estimators. OLS are inappropriate as they would result in biased estimates on CS data and inconsistent estimates on TS and panel data (Pesaran and Smith, 1995). It is therefore necessary to apply other techniques, such as GLS or ML estimation. Another approach, suggested in Pesaran and Smith (1995) and referred to as the "mean group estimator", consists in estimating on (large) panel data separate regressions for each group and then averaging the coefficients over groups. The related standard errors will have to be calculated explicitly, as dynamic TS models usually provide incorrect standard errors. This procedure will result in *efficient* estimates – i.e. unbiased and with a minimal variance.

A.3. Estimation methods

Assume that we would like to estimate the following simple linear regression model:

$$y_t = x_t\beta - u_t{}^{10}$$

This model describes a linear relationship between a given variable y (the one we want to explain, i.e. the dependent variable) and other variables that are supposed to affect the dependent variable y. These other variables are referred to as the *explanatory variables* (unpredetermined variables). u_t represents an error term that encompasses everything that has not been observed, and thus what is unknown. u_t is assumed to satisfy the following conditions:

$$E(u_t) = 0$$

$$V(u_t) = \sigma^2 \text{ for all } t$$

$$E(u_t u_s) = 0 \text{ for all } t \neq s$$

The matrix X, defined such as $X = (x_1, x_2, ..., x_N)$, gathers all the explanatory variables that are supposed to affect the dependent variable y. β is the parameter to be estimated to identify the relationship between y and X.

The usual way to estimate this model is to apply the ordinary least squares (OLS) method. Let $\hat{\beta}$ be the estimate of β by OLS. Given $\hat{\beta}$, one can define y (the predicted values of y) as:

$$\hat{y}_t = x_t\hat{\beta} + u_t$$

One can also define $\hat{u}$ as: $\hat{u}_t = y_t - x_t\hat{\beta} = y_t - \hat{y}$. $\hat{u}_t$ is the vector of residuals measuring the deviation that exist between the observed values of y (y) and the predicted values of y ($\hat{y}$). From this, one can define the sum of squared residuals as $\Sigma u_t^2 = u'u = (y - X\beta)'(y - X\beta)$.

The OLS approach consists in finding the value $\hat{\beta}$ for β where the sum of squared residuals is minimised: $OLS \rightarrow \hat{\beta} / \min_{\beta} u'u$.

Under stationarity, the OLS estimator is unbiased and has a smaller variance than other estimators of the linear regression model.[11] When an estimator combines these two properties, it is said to be efficient.

In some cases, the assumptions concerning the error term u are not satisfied. For example, the error terms might be correlated with themselves (autocorrelation) or the variance of the error term might not be constant over time (heteroscedasticity), such that $V(u_t) = \sigma_t^2$ instead of $V(u_t) = \sigma^2$. In these circumstances, the OLS approach should not be applied, as it would provide biased and inconsistent estimates of *β*. In these contexts, it is of common practice to apply what is known as the generalised least squares (GLS) method. The GLS approach consists in applying the OLS method on weighted data. Data are weighted by a matrix H such as $H'H = [V(u)]^{-1} = [E(uu')]^{-1}$. Therefore, the transformation leads to the following regression model:

$$y = X\beta + u \Rightarrow Hy = HX\beta + Hu \Rightarrow y^* = X^*\beta + u^*$$

This transformation makes the weighted u (u^*) homoscedastic (i.e. they have a variance constant over time) and therefore, OLS can be applied.

The GLS estimator of *β* will be more consistent and more efficient than the OLS estimator in presence of heteroscedasticity or autocorrelation.

Another well known problem is when the explanatory variables X (also referred to as "regressors") and the error terms are correlated, *i.e.* there is *serial correlation*. In this context, a certain amount of the variation in the dependent variable y due to the variation in u is attributed to the explanatory variables because of their correlation with u. If OLS are applied in presence of serial correlation, the estimates would be biased and inconsistent. It is then recommended to apply another methodology called *instrumental variables* (IV) or *instruments*.

The IV approach consists in building variables (the *instruments*) that will substitute the X that are correlated with u. This methodology generally requires two steps. In the first step, the instruments W are created. In the second step, the correlated explanatory variables X are replaced by W. It is then possible to estimate the regression model by OLS. The choice of the instruments W is crucial. When the IV approach is used with TS data, it is of common practice to take lagged values of the correlated regressors as instruments. The IV estimator is consistent and asymptotically[12] Gaussian (*i.e.* it asymptotically follows the Normal distribution). However, as it can be biased in small samples, it is only recommended to use an IV estimator in large samples.

These three methodologies are usually applied when one does not know the distribution function that has generated the sample from which the data come from. However, when the specific form of the probability distribution function is known, it is possible to obtain estimates more efficient than OLS, GLS or IV estimates. More accurate and more appropriate estimates can be obtained through the use of the maximum likelihood (ML) approach.

The ML approach consists in the maximisation of the likelihood function $L(\theta/y)$ in order to find the particular value of θ that is the most likely to have generated the sample y. The likelihood function $L(\theta/y)$ is the joint density function of the observation of the sample y: $L(\theta/y) = f(y/\theta)$, assuming independent data. The joint density function describes how the probability density varies when the sample changes according to a specific parameter θ. $L(\theta/y)$ can be therefore defined as a measure of the likelihood that the sample y has been generated by the joint density function characterised by θ. The ML estimator has some interesting properties:

- "When applied to small samples, the ML estimator (MLE) is unbiased and has small variance (*i.e.* it is efficient). In addition, it is invariant under reparameterisation: if $\hat{\theta}$ is the MLE of θ and $g(\theta)$ is a continuous function of θ, then $g(\hat{\theta})$ is the MLE of $g(\theta)$.

- "When applied to large samples, the MLE is consistent, asymptotically normal and asymptotically efficient.

A.4. Conclusion

There is no consensus on the most appropriate type of data and estimation approach to estimate price elasticities. Several estimation approaches can be valid according to the type of data, the type of elasticity requested (short term or long term), the dynamic process chosen, the characteristics of the demand function, etc. If one wants to estimate short term elasticities, estimation of cross sectional, time series or panel data by OLS can provide similar mean estimates of the coefficients. However, it becomes more complicated when dealing with the estimation of long term elasticities which requires the use of dynamic models. The above considerations suggest that estimation of time series data, according to Pesaran and Smith's technique (1995), will provide efficient long term estimates of price elasticities. If time series data are not available, cross sections can provide meaningful point estimates of long term elasticities, and they will in general be more robust to misspecification than time series data.

Notes

1. For more details on the different econometric techniques and description of panel data models, see Greene (1997).

2. Cross-sectional data are not concerned by integration issues as they do not explicitly take time into account.

3. A brief presentation of the cointegration techniques is proposed in the next section. For more details on non-stationary time series and cointegration, see Greene (1997).

4. An error term u is considered as a *white noise* when it satisfies the following assumptions: the mean, or expected value, of u is zero (formally $E(u_i) = 0$ for all i), the variance of u is σ^2 (formally $E(u_i^2) = \sigma^2$ for all i), and the covariance between u_i and u_j is zero (formally $E(u_iu_j) = 0$ for $i \neq j$), i.e u_i and u_j are not correlated.

5. The causes for such correlation include measurement errors or endogeneity of some explanatory variables. For more details, see Greene (1997).

6. For more details, see Greene (1997).

7. An estimator for a parameter b is said to be consistent if the estimator converges in probability to the true value of the parameter β. Formally, $\text{plim}(b) \longrightarrow \beta$.

8. An estimator b of a distribution's parameter β is unbiased if the mean of b's sampling distribution is β. Formally, b is unbiased if $E(b) = \beta$ when $N \longrightarrow \infty$.

9. A variable is said to be endogenous in a model if it is, at least partly, a function of other parameters and variables in the model.

10. This can be also expressed in the vector form as $y = X\beta - u$.

11. The OLS estimator is said to be "the best linear unbiased estimator" (BLUE).

12. "Asymptotically" refers to large sample properties.

References

AASNESS, Jørgen and Erling RØED LARSEN (2002), *Distributional and Environmental effects of Taxes on Transportation*, Discussion Papers No. 321, Statistics Norway. Available at *www.ssb.no/publikasjoner/DP/pdf/dp-321.pdf*.

AGNOLUCCI, Paolo and Paul EKINS (2004), *The Announcement Effect and Environmental Taxation*. Tyndall Centre for Climate Change Research, Working Paper 53, University of East Anglia, Norwich. Available at *www.tyndall.ac.uk/publications/working_papers/wp53.pdf*.

AGNOLUCCI, Paolo, Terry BARKER and Paul EKINS (2004), *Hysteresis and Energy Demand: The Announcement Effects and the Effects of the UK Climate Change Levy*. London: Policy Studies Institute, mimeo; and Tyndall Centre for Climate Change Research, Working Paper 51, University of East Anglia, Norwich. Available at *www.tyndall.ac.uk/publications/working_papers/wp51.pdf*.

ANDERSEN, Mikael Skou (1999), "Governance by green taxes: implementing clean water policies in Europe 1970-1990", *Environmental Economics and Policy Studies* (1999) Vol. 2, pp. 39-63.

ARMINGTON, Paul (1969), "A theory of demand for products distinguished by place of production", *IMF Staff papers*. IMF, Washington DC.

ASHIABOR, Hope, Kurt DEKETELAERE, Larry KREISER and Janet MILNE (eds.) (2005), *Critical Issues in Environmental Taxation: International and Comparative Perspectives: Vol. II*, Richmond Law and Tax Ltd, Richmond, UK.

ASSUNÇÃO, Lucas and Xiang Zhong ZHANG (2002), *Domestic Climate Change Policies and the WTO*, United Nations Conference on Trade and Development, UNCTAD Discussion Papers, No. 164, Geneva, November 2002.

BACH, Stephan *et al.* (2001), *Die ökologische Steuerreform in Deutschland* (The ecological tax reform in Germany), (In German). Physika-Verlag, Heidelberg 2001.

BACH, Stephan (2005), *Be- und Entlastungswirkungen der Ökologischen Steuerreform nach Produktionsbereichen* (Negative and positive impacts on prodution sectors of the ecological tax reform), Report prepared for the Federal Environment Agency, Berlin. (In German.) Available at *www.umweltbundesamt.org/fpdf-l/2960.pdf*.

BARANZINI, Andrea and Philippe THALMANN (eds.), *Voluntary Approaches in Climate Policy*. Edward Elgar, Cheltenham.

BARDE, Jean-Philippe and Nils Axel BRAATHEN (2005), "Environmentally Related Levies" in Sijbren CNOSSEN (ed.) (2005) *Theory and Practice of Excise Taxation*, Oxford University Press, Oxford, United Kingdom.

BARDE, Jean-Philippe and Nils Axel BRAATHEN (forthcoming), "Green Tax Reforms in OECD Countries: An Overview". In Peter N. NEMETZ (ed.) (forthcoming), *Sustainable Resource Management: Reality or Illusion?* Edward Elgar, Cheltenham.

BARKER, Terry, Paul EKINS and Nick JOHNSTONE (eds.) (1995), *Global Warming and Energy Demand*, Routledge, 1995.

BARTELINGS, Helen *et al.* (2005), *Effectiveness of landfill taxation*, Report prepared for the Dutch Ministry of Housing, Spatial Planning and the Environment. Institute for Environmental Studies, Vrije Universiteit, Amsterdam. Available at *www.ivm.falw.vu.nl/Research_output/index.cfm/home_subsection.cfm/subsectionid/FF91BCBD-EAFE-426A-ABB8184073A39BBF*.

BATES, J. (1995), *Full fuel Cycle Atmospheric Emissions and Global Warming Impacts from UK Electricity Generation*. Energy Technology Support Unit, Harwell.

BENEDICK, Richard Elliott (1998), *Ozone Diplomacy: New Directions in Safeguarding the Planet*, Enlarged edition, Cambridge, MA, Harvard University Press, 1998.

BIERMANN, Frank and BROHM, Rainer (2003), *Implementing the Kyoto Protocol Without the United States: The Strategic Role of Energy Tax Adjustments at the Border*, Global Governance Working Paper No. 5, January 2003, available at: *http://glogov.org*.

BJØRNER, Thomas Bue and Henrik Holm JENSEN (2002), "Energy taxes, voluntary agreements and investment subsidies – a micro-panel analysis of the effect on Danish industrial companies' energy demand". *Resource and Energy Economics*, Vol. 24 (2002), pp. 229-249.

BJERTNÆS, Geir H. and Taran FÆHN (2004), *Energy Taxation in a Small, Open Economy: Efficiency Gains under Political Restraints*. Discussion Papers No. 387, Statistics Norway. Available at *www.ssb.no/publikasjoner/DP/pdf/dp387.pdf*.

BOLTHO, A. (1996), "The assessment: international competitiveness", *Oxford Review of Economic Policy*, Vol. 12, No. 3, pp. 1-16.

BORK, Christhart (2003), *Distributional Effects of the Ecological Tax Reform in Germany – An Evaluation with a Microsimulation Model*. Paper for the OECD workshop on the Distribution of Benefits and Costs of Environmental Policies, Paris, 4-5 March 2003. In SERRET and JOHNSTONE (eds.) (2006), *The Distributional Effects of Environmental Policy*, Edward Elgar, Cheltenham and OECD, Paris.

BOSELLO, F., C. CARRARO and M. GALEOTTI (2001), "The Double Dividend Issue: Modeling Strategies and Empirical Findings", *Environment and Development Economics*, Vol. 6, pp. 9-45.

BOVENBERG, A. Lans (1995), "Environmental Taxation and Employment", *De Economist 143*, pp. 111-140.

BOVENBERG, A. Lans (1998), "Environmental Taxation and the Double Dividend"*Empirica 25*, pp. 15-35.

BOVENBERG, A. Lans (1999), "Green Tax Reforms and the Double Dividend: An Updated Reader's Guide", *International Tax and Public Finance 6*, pp. 421-443.

BOVENBERG, A. Lans and Lawrence H. GOULDER (2003), *Confronting industry-distributional concerns in U.S. climate-change policy*. Les séminaires de l'Iddri, No. 6. Institut du Développement Durable et des Relations Internationales. Available at *www.iddri.org/iddri/telecharge/mardis/06_goulder.pdf*.

BOVENBERG, A.L. and F. van der PLOEG (1994), *Environmental policy, public finance and the labour market in a second-best world*, Journal of Public Economics, 55, pp. 349-390.

BRAATHEN, Nils Axel (2005), "Environmental Agreements Used in Combination with Other Policy Instruments", in Edoardo CROCI (ed.) (2005), *The Handbook of Environmental Voluntary Agreements*, Springer, Dordrecht.

BRACK, Duncan (1996), *International Trade and the Montreal Protocol*, The Royal Institute of International Affairs, London, Earthscan.

BRACK, Duncan (1997), *The Growth and Control of Illegal Trade in Ozone-Depleting Substances*. Paper delivered at the 1997 Taipei International Conference on Ozone Layer Protection (December, 1997).

BRACK, Duncan, Michael GRUBB and Craig WINDRAM (2000), *International Trade and Climate Change Policies*, The Royal Institute of International Affairs, London, Earthscan.

BRÄNNLUND, Runar and Jonas NORDSTRÖM (2004), *Carbon Tax Simulations Using a Household Demand Model*. European Economic Review 48, 2004, pp. 311-333.

BRUVOLL, Annegrete and Bodil Merete LARSEN (2004), "Greenhouse gas emissions in Norway: Do carbon taxes work?"*Energy Policy 32* (2004). A previous version issued as Discussion Papers 337 by Statistics Norway, Oslo is available at *www.ssb.no/cgi-bin/publsoek?job=forside&id=dp-337&kode=dp&lang=en*.

BYE, Torstein, Maria Wist LANGMOEN and Jørgen AASNESS (2004), *Pris- og inntektselastisiteter for husholdningenes etterspørsel etter elektrisitet – en metaanalyse for nordiske land*. (Price- and income-elasticities for households' demand for electricity – a meta-analysis for Nordic countries.) (In Norwegian.) Unpublished paper. Statistics Norway, Oslo.

Cambridge Econometrics (2005), *Modelling the Initial Effects of the Climate Change Levy*. Report submitted to HM Customs and Excise by Cambridge Econometrics, Department of Applied Economics, University of Cambridge and the Policy Studies Institute. Available at *http://customs.hmrc.gov.uk/channelsPortalWebAppchannelsPortalWebApp.portal?_nfpb=true&_pageLabel=pageLibrary_Miscellaneous Reports&propertyType=document&columns=1&id=HMCE_PROD1_023971*.

CAMERON, J. (1993), "The GATT and the Environment," in P. SANDS (ed.), *Greening International Law*, London, Earthscan, 110, p. 113.

Carbon Trust (2004), *The European Emissions Trading Scheme: Implications for Industrial Competitiveness*. The Carbon Trust, London. Available at *www.carbontrust.co.uk/Publications/CT-2004-04.pdf*.

CEC (Commission of the European Communities) (2001a). *Proposal for a Directive of the European Parliament and the Council Establishing a Framework for the GHG Emissions Trading within the European Community and Amending Council Directive 96/61/EC.* COM(2001)581 Final – 2001/0245 (COD). Commission of the European Communities, Brussels. Available at *http://europa.eu.int/eur-lex/en/com/pdf/2001/en_501PC0581.pdf.*

CEC (2002a), *Fiscal Measures to Reduce CO_2 Emissions from New Passenger Cars.* Report prepared by COWI for the European Commission, DG Environment, Brussels. Available at *http://europa.eu.int/comm/taxation_customs/resources/documents/co2_cars_study_25-02-2002.pdf.*

CEC (2002b), *Amended Proposal for a Directive of the European Parliament and the Council Establishing a Framework for the GHG Emissions Trading within the European Community and Amending Council Directive 96/61/EC.* COM(2002)680 Final – 2001/0245 (COD). Commission of the European Communities, Brussels. Available at *http://europa.eu.int/eur-lex/en/com/pdf/2002/com2002_0680en01.pdf.*

CEC (2002c), *European Agriculture Entering the 21th Century.* European Commission, Directorate General for Agriculture, Brussels.

CEC (2005), *Proposal for a Council Directive on passenger car related taxes,* COM(2005)261 final. Commission of the European Communities, Brussels. Available at *http://europa.eu.int/comm/environment/co2/pdf/taxation_com_2005_261.pdf.*

CEU (Council of the European Union) (1999), *Council Directive 1999/31/EC of 26 April 1999 on the landfill of waste,* CEU, Brussels. Available at *http://europa.eu.int/eur-lex/pri/en/oj/dat/1999/l_182/l_18219990716en00010019.pdf.*

CEU (2003), *Council Directive 2003/96/EC of 27 October 2003 restructuring the Community framework for the taxation of energy products and electricity.* Council of the European Union, Brussels. Available at *http://europa.eu.int/eur-lex/pri/en/oj/dat/2003/l_283/l_28320031031en00510070.pdf.*

CHARNOVITZ, Steve (1993), "Environmental trade measures and economic competitiveness: an overview of the issues" in OECD (1993), *Environmental Policies and Industrial Competitiveness,* OECD, Paris, pp. 141-9.

CHARNOVITZ, Steve (2003), *Trade and Climate: Potential Conflicts and Synergies,* The Pew Centre on Global Climate Change, Working Draft, July 2003.

CNOSSEN, Sijbren (ed.) (2005), *Theory and Practice of Excise Taxation,* Oxford University Press, Oxford, United Kingdom.

Commission investigating the competitive conditions in air traffic (1999), *Om betydningen av skatter, avgifter og gebyrer for luftfartens konkurransevilkår* (The significance of taxes, duties and fees for the competitive conditions in air traffic). Report in Norwegian, prepared for the Ministry of Finance. Oslo.

Conseil des impôts (2005), *Fiscalité et environnement. Vingt-troisième rapport au Président de la République.* Conseil des impôts, Paris. Available at *www.ccomptes.fr/organismes/conseil-des-impots/rapports/fiscalite-environnement/rapport.pdf.*

CONVERY, Frank, Simon McDONNELL and SUSANA FERREIRA (2005), *The Most Popular Tax in Europe? Lessons from the Irish Plastic Bags Levy.* Paper submitted to the EAERE conference, Bremen, 23-26 June 2005.

COOK, Elizabeth (ed.) (1996), *Ozone Protection in the United States: Elements of Success,* World Resources Institute.

CORNILLIE, J. (2003), *Developments in EU CO_2 Emissions Allowance Trading.* Paper to OXERA Environmental Policy Group. Oxera, Oxford.

CORNWELL, Antonia and John CREEDY (1997), "Measuring the Welfare Effects of Tax Changes Using the LES: An Application to a Carbon Tax". *Empirical Economics,* 22, pp. 589-613.

CROCI, Edoardo (ed.) (2005), *The Handbook of Environmental Voluntary Agreements,* Springer, Dordrecht.

Danish Commerce and Companies Agency (2005), *Aktivitetsbaseret måling af virksomheders administrative byrder af erhvervsrelatert regulering på Skatteministeriets område.* (Activity-based measurement of the administrative burden of business-related regulation under the responsibility of the Ministry of Taxation). Danish Commerce and Companies Agency, Copenhagen. (In Danish, with English summary.) Available at *www.amvab.dk/graphics/AdmLet/Publikationer/AMVAB_SKM_hovedrapport.pdf.*

DAVIE, Bruce F. (1995), *Border Tax Adjustments for Environmental Excise Taxes: The US Experience.* Paper presented for the Allied Social Science Associations, January 1995, Washington.

DAVIES, Bob and Michael DOBLE (2004), "The development and implementation of a Landfill Tax in the UK", in OECD (2004a), *Addressing the Economics of Waste*, OECD, Paris. Available at *http://www.1.oecd.org/publications/e-book/9704031E.PDF.*

DAVIES, Bob and Helen DUNN (2003), Contribution prepared by the United Kingdom for the OECD Workshop on the distribution of Benefits and Costs of Environmental Policies: Analysis, Evidence and Policy Issues, Paris, 4-5 March 2003.

DEFRA (Department for the Environment, Food and Rural Affairs) (2001), *Regulatory and Environmental Impact Assessment of Proposal to Introduce the Climate Change Agreements (Eligible Facilities) Regulations 2001*, DEFRA, London Available at *www.defra.gov.uk/environment/ccl/reia-662/index.htm.*

DEFRA (2004a), *Review of Environmental and Health Effects of Waste Management: Municipal Solid Waste and Similar Wastes*, DEFRA, London. Available at *www.defra.gov.uk/environment/waste/research/health/pdf/health-report.pdf.*

DEFRA (2004b), *Review of Environmental and Health Effects of Waste Management: Municipal Solid Waste and Similar Wastes. Extended Summary*, DEFRA, London. Available at *www.defra.gov.uk/environment/waste/research/health/pdf/health-summary.pdf.*

DEFRA (2004c), *Economic Valuation of the External Costs and Benefits to Health and Environment of Waste Management Options*, DEFRA, London. Available at *www.defra.gov.uk/environment/waste/research/health/pdf/costbenefit-valuation.pdf.*

DELACHE, Xavier (2002), "Comments on the Discussion Paper" in OECD (2002b), *Implementing Environmental Fiscal Reform: Income Distribution and Sectoral Competitiveness Issues.* Proceedings of a Conference held in Berlin, Germany, 27 June 2002. OECD, Paris.

DEMARET, Paul and Raoul STEWARDSON (1994), *Border Tax Adjustments Under GATT and EC Law and General Implications for Environmental Taxes.* 28, Journal of World Trade 4.

DETR (Department of the Environment, Transport and the Regions) (2001), *Tradable landfill permits consultation paper.* DETR, London. Available at *www.defra.gov.uk/environment/consult/tradeperm/pdf/tradable.pdf.*

Deutscher Bundestag (2002), *Antwort der Bundesregierung* (Answers from the Federal Government), Drucksache 14/9993, 7.10.2002, Deutscher Bundestag, Berlin. Available at *http://dip.bundestag.de/btd/14/099/1409993.pdf.*

DIJKGRAAF, Elbert (2004), *Regulating the Dutch Waste Market*, Ph.D. thesis, Research Centre for Economic Policy, Erasmus University, Rotterdam. The conclusions and a summary are available at *www.seor.nl/ecri/pdf/psdef_samenv.pdf.*

DIJKGRAAF, Elbert and Herman R.J. VOLLEBERGH (2004), "Burn or bury? A social cost comparison of final waste disposal methods". *Ecological Economics, 50 (2004).*

DORNBUSH, R. and J.M. POTERBA (eds.) (1991), *Global Warming: Economic Policy Responses to Global Warming*, The MIT Press, Cambridge, Mass.

DTI (Department of Trade and Industry) and DEFRA (2004), *Creating a Low Carbon Economy: First Annual Report on Implementation of the Energy White Paper.* DTI, London.

ECMT (European Conference of Ministers of Transport) (2000), *Efficient Transport Taxes & Charges*, OECD/ECMT, Paris.

ECMT (2003), *Reforming Transport Taxes*, OECD/ECMT, Paris.

ECOFYS (2000), *Greenhouse gas emissions from major industrial sources – III Iron and Steel Production.* Report no. PH3/30, IEA Greenhouse Gas R&D Programme. OECD/IEA, Paris.

ECON (2000), *Miljøkostnader ved avfallsbehandling* (Environmental costs of waste treatment), Report 85/00. (In Norwegian). Econ, Oslo.

ECOTEC (1999), *Who Gains from the Climate Change Levy?* Report to WWF UK. ECOTEC, Birmingham. Available at *www.wwf.org.uk/filelibrary/pdf/whogains.pdf.*

EDERINGTON, Josh, Arik LEVINSON and Jenny MINIER (2003), *Footloose and pollution free.* Working Paper 9718, National Bureau of Economic Research, Washington DC.

EEA (European Environment Agency) (2005), *Market-based instruments for environmental policy in Europe*, EEA Technical Report 8/2005. European Environment Agency, Copenhagen. Available at *http://reports.eea.eu.int/technical_report_2005_8/en/EEA_technical_report_8_2005.pdf.*

EEA (2006), *Transport and environment: facing a dilemma*. EEA Report No. 3/2006. European Environment Agency, Copenhagen. Available at *http://reports.eea.eu.int/eea_report_2006_3/en/term_2005.pdf*.

EKINS, Paul and Simon DRESNER (2004), *Green Taxes and Charges – Reducing their impact on low-income households*, Joseph Rowntree Foundation, York.

Energy Policy (2006), "Social and political responses to ecological tax reform in Europe". *Energy Policy*, Volume 34, No. 8, pp. 895-970.

ETSU (Energy Technology Support Unit) (2001), *Climate Change Agreements – Sectoral Energy Efficiency Targets*. Available at *www.defra.gov.uk/environment/ccl/pdf/etsu-analysis.pdf*.

ETSU Future Energy Solutions (2003), *Climate Change Agreements – Results of the First Target Period Assessment. Version 1.1 – Preliminary Results*. Available at *www.defra.gov.uk/environment/ccl/pdf/cca_tp1_prelim.pdf*.

EU (European Union) (2003a), *Directive 2003/87/EC of the European Parliament and of the Council of 13 October 2003 establishing a scheme for greenhouse gas emission allowance trading within the Community and amending Council Directive 96/61/EC*. EU, Brussels. Available at *http://europa.eu.int/eur-lex/pri/en/oj/dat/2003/l_275/l_27520031025en00320046.pdf*.

Eurostat (2002), Agriculture in the European Union – Statistical and economic information. Eurostat, Luxembourg.

Eurostat (2003), *Energy Taxes in the Nordic Countries – Does the polluter pay?* Report prepared for Eurostat by National Statistical offices in Norway, Sweden, Finland and Denmark. Eurostat, Luxembourg. Available at *www.scb.se/statistik/MI/MI1202/2004A01/MI1202_2004A01_BR_MIFT0404.pdf*.

FAUCHALD, Ole Kristian (1998), *Environmental Taxes and Trade Discrimination*, Kluwer Law, The Hague.

FLOYD, Rober, H. (1973), "GATT Provisions on Border Tax Adjustments", (1979) *Journal of World Trade Law*, 489.

FULLERTON, Don (2005), "An Excise Tax on Municipal Solid Waste?" in Sijbren CNOSSEN (ed.) (2005) *Theory and Practice of Excise Taxation*, Oxford University Press, Oxford, United Kingdom.

FULLERTON, Don and Garth HEUTEL (2005), *The General Equilibrium Incidence of Environmental Taxes*, Working Paper 11311, National Bureau of Economic Research. Cambridge, Massachusetts. Available at *www.eco.utexas.edu/faculty/Fullerton/papers/fh-taxes.pdf*.

FULLERTON, Don and Gilbert METCALF (2001), "Environmental Controls, Scarcity Rents and Pre-existing Distortions". *Journal of Public Economics* 80 (2001), pp. 249-267. Available at *www.eco.utexas.edu/faculty/Fullerton/papers/fm-jpube01.pdf*.

GATT (1970), *Border Tax Adjustments, Report of the Working Party*, L/3464, adopted 2 December 1970.

GATT (1976), *United States – Domestic International Sales Corporation Scheme*, GATT BISD, 23S/98 (1977), adopted 2 November 1976.

GATT (1986), *Committee on Subsidies and Countervailing Measures*, Minutes of the Meeting held on 22-23 April 1986, SCM/M/31.

GATT (1987), *United States – Taxes on Petroleum and Certain Imported Substances, [Superfund] Report of the Panel*, GATT Doc. L/6175, BISD 34S/36, adopted 17 June 1987.

GATT (1991), *United States – Restrictions on Imports of Tuna, Report of the Panel*, Geneva: GATT, DS21/R.

GATT (1996), *Japan – Customs Duties Taxes and Labelling Practices on Imported Wines and Alcoholic Beverages*, Report of the Appellate Body, WT/DS8/AB/R.

GATT (1992), *United States – Measures Affecting Alcoholic and Malt Beverages*, Report of the Panel, GATT Document DS23/R, GATT BISD 39S/206; adopted 19 June 1992.

GATT (1994), *United States – Restrictions on Imports of Tuna, Report of the Panel*, Geneva: GATT, DS29/R.

GAVERUD, Henrik (2004), *Benefits from Environmental Taxation: A Case Study of the US Tax on Ozone Depleting Substances*, Masters thesis (2004:035 SHU), Luleå University of Technology, Department of Business Administration and Social Sciences.

German Council for Sustainable Development (2001), *Ziele zur nachhaltigen Entwicklung in Deutschland* (Targets for Sustainable Development in Germany) (in German). German Council for Sustainable Development, Berlin. Available at: *www.nachhaltigkeitsrat.de/service/download/pdf/RNE_Dialog-papier.pdf*.

GILLINGHAM, Kenneth, Richard NEWELL, and Karen PALMER (2004), *Retrospective Examination of Demand-Side Energy Efficiency Policies*. Discussion Paper 04-19 rev, Resources for the Future, Washington DC. Available at *www.rff.org/Documents/RFF-DP-04-19rev.pdf*.

GLAISTER, Stephen and Daniel J. GRAHAM (2004), *Pricing our Roads: Vision or Reality?* The Institute of Economic Affairs, London.

GOH, Gavin (2004), "The World Trade Organization, Kyoto and Energy Tax Adjustments at the Border", 38 (3) *Journal of World Trade*: 395

GOODWIN, Phil, Joyce DARGAY and Mark HANLY (2004), "Elasticities of Road Traffic and Fuel Consumption with Respect to Price and Income: A Review". *Transport Reviews,* Vol. 24, No. 3, pp. 275-292. Available at *www.cts.ucl.ac.uk/tsu/papers/transprev243.pdf.*

GOULDER, Lawrence H. (1995), "Environmental Taxation and the 'Double Dividend': A Reader's Guide", *International Tax and Public Finance* 2, pp. 157-184.

GOULDER, Lawrence H., Ian W. H. PARRY and Dallas BURTRAW (1997), *Revenue-Raising versus Other Approaches to Environmental Protection: The Critical Significance of Preexisting Tax Distortions.* Rand Journal of Economics, Vol. 28, No. 4, pp. 708-731. Available at *www.rje.org/abstracts/abstracts/1997/ Winter_1997._pp._708_731.html.*

Government of Japan (1996) "*Utilization of Economic Instruments in Environmental Policies – Taxes and Charges*", First Report of the Research Panel on Economic Instruments such as Taxation and Charges in Environmental Policies, Environment Agency, Government of Japan. Available at: *www.env.go.jp/en/rep/etax/et2e.html.*

GRAHAM, Daniel J. and Stephen GLAISTER (2002), *Review of Income and Price Elasticities of Demand for Road Traffic,* Final Report to the Department of Transport, London, UK, July 2002. Downloadable at: *www.cts.cv.imperial.ac.uk/documents/publications/iccts00267.pdf.*

Green Budget Germany (2004), *Ecotaxes in Germany and the United Kingdom – A Business View.* Report on a Conference hosted by Green Budget Germany in cooperation with The Heinrich Böll Foundation and the Anglo-German Foundation. Available at *www.eco-tax.info/downloads/ConferenceReport.pdf.*

GREENE W.H. (1997), *Econometric Analysis,* Third Edition, Prentice-Hall, United-States.

HAMILTON, James T. and W. Kip VISCUSI (1999), *Calculating Risks? The Spatial and Political Dimensions of Hazardous Waste Policy.* Cambridge, Mass. MIT Press.

HAMILTON, James T. (2003), *Environmental Equity and the Siting of Hazardous Waste Facilities in OECD Countries: Evidence and Policies.* Paper prepared for the OECD Workshop on the Distribution of Benefits and Costs of Environmental Policies, Paris, 4-5 March 2003. Will be included in SERRET and JOHNSTONE (eds) (2006), *The Distributional Effects of Environmental Policy,* Edward Elgar, Cheltenham and OECD, Paris.

HANLY M., J. DARGAY and P. GOODWIN (2002), *Review of Income and Price Elasticities in the Demand for Road Traffic,* Final Report to the Department of Transport, Local Government and the Regions (DTLR) (now Department of Transport), London, UK, March 2002. Available at: *www.dft.gov.uk/ stellent/groups/dft_econappr/documents/downloadable/dft_econappr_033848.pdf.*

HAUTZINGER, Heintz et al. (2004), *Analyse von Änderungen des Mobilitätsverhaltens – insbesondere der Pkw-Fahrleistung – als Reaktion auf geänderte Kraftstoffpreise.* (Analysis of changes in mobility conditions – in particular for passenger transport – in response to changed fuels prices.) (In German.) Final report of a research project undertaken for the Federal Ministry of Transport, Germany, Bonn. Available at *www.ivt-verkehrsforschung.de/pdf/Kraftstoffpreise_und_Mobilitaet.pdf.*

HARRISON, David (1999), "Tradable Permits for Air Pollution Control: The United States Experience" in OECD (1999b), *Implementing Domestic Tradable Permits for Environmental Protection.* OECD Proceedings, OECD, Paris.

HAYLER, J. (2003), *The UK Climate Change Levy: an Event Study.* Master's Thesis, Environmental Economics, University College London.

HENKENS, P.L.C.M. and H. van KEULEN (2001), "Mineral policy in the Netherlands and nitrate policy within the European Community". *Netherlands Journal of Agricultural Science.* 49: pp. 117-134.

HERTEL T., D. HUMMELS, M. IVANIC and R. KEENEY (2004), "How Confident Can We Be in CGE-Based Assessments of Free Trade Agreements?", *NBER Working Paper* 10477. Available at: *www.nber.org/ papers/w10477.pdf.*

HM Customs and Excise (1999), *Budget 99: A Climate Change Levy,* HM Customs and Excise, London.

HM Customs and Excise (2004), Combining the government's two health and environment studies to calculate estimates for the external costs of landfill and incineration, HM Customs and Excise, London. Available at *http://customs.hmrc.gov.uk/channelsPortalWebApp/channels PortalWebApp.portal?_nfpb=true&_pageLabel=pageVAT_ShowContent&id=HMCE_PROD_011566&propert yType=document.*

HM Treasury (1997), *Tax Measures to Help the Environment*. New Release, July 2. HM Treasury, London.

HM Treasury (2003), *The Green Book: Appraisal and Evaluation in Central Government*, HM Treasury, London, Available at *www.hm-treasury.gov.uk/media/785/27/Green_Book_03.pdf*.

HM Treasury (2005), *Budget 2005, Chapter 7 – Protecting the environment*. HM Treasury, London. Available at *www.hm-treasury.gov.uk/media/AA7/59/bud05_chap07_171.pdf*.

HM Treasury (2006), *Budget 2006, Chapter 7 – Protecting the environment*. HM Treasury, London. Available at *www.hm-treasury.gov.uk/media/20F/1D/bud06_ch7_161.pdf*.

HOERNER, A.ndrew J. and B. BOSQUET (2001) (1997), *Alternative Approaches to Offsetting the Competitive Burden of a Carbon/Energy Tax*, Environmental Tax Program, August 1997.

HOERNER, Andrew J. (1998), *The Role of Border Tax Adjustments in Environmental Taxation: Theory and US Experience*, presented at the International Workshop on Market Based Instruments and International Trade of the Institute for Environmental Studies, Amsterdam, the Netherlands, 19 March 1998.

HOERNER, Andrew J. and Benoit BOSQUET (2001), *Environmental Tax Reform: The European Experience.*, Center for a Sustainable Economy, Washington DC., February 2001.

HOERNER, Andrew J. and Frank MULLER (1996), *Carbon Taxes for Climate Protection in a Competitive World*, a paper prepared for the Swiss Federal Office for Foreign Economic Affairs by the Environmental Tax Program of the Center for Global Change, University of Maryland College Park (June 1996).

HOEVENAGEL, Ruud, Edwin van NOORT and Rene de KOK (1999), *Study on a European Union wide regulatory framework for levies on pesticides*. Report commissioned by the European Commission, DG XI. Available at *http://europa.eu.int/comm/environment/enveco/taxation/eimstudy.pdf*.

HOLMØY, Erling (2005), *The Anatomy of Electricity Demand: A CGE Decomposition for Norway*. Discussion Paper 426, Statistics Norway, Oslo. Available at *www.ssb.no/publikasjoner/DP/pdf/dp426.pdf*.

HOURCADE, J.-C. and P. SHUKLA (2001), "Global, Regional, and National Costs and Ancillary Benefits of Mitigation", Chapter 8 in *Climate Change 2001*, IPCC Third Assessment Report, Cambridge University Press.

HUFBAUER, G.C. (1993), *The Evolution of Border Tax Adjustments*, Report prepared for the Center for Strategic Tax Reform, March 1993.

HÖGLUND ISAKSSON, Lena (2005), "Abatement costs in response to the Swedish charge on nitrogen oxide emissions". *Journal of Environmental Economics and Management*, 50, 102-120.

IEA (International Energy Agency) (1999), *The reduction of greenhouse gas emission from the cement industry*. IEA Greenhouse Gas R&D Programme. IEA/OECD, Paris.

JACOBY, Henry D. and A. Denny ELLERMAN (2004), "The safety valve and climate policy", *Energy Policy*, 32.

JAFFE, A.B., S.R. PETERSON, P.R. PORTNEY and R.N. STAVINS (1995), "Environmental regulation and the competitiveness of US manufacturing: what does the evidence tell us?", *Journal of Economic Literature*, Vol. 33, pp. 132-163.

JÄNICKE, Martin, Lutz MEZ, Pernille BECHSGAARD and Børge KLEMMENSEN (1998), *Innovation and Diffusion through Environmental Regulation: The Case of Danish Refrigerators*. FFU-report 98-3, Forschungsstelle für Umweltpolitikk, Freie Universität Berlin. Available at *http://web.fu-berlin.de/ffu/ffu_e/index.htm*.

Japanese Environment Agency (1997), "Japan Economic Quarterly: News from the Environment Agency", September 1997.

JOHNSTONE, Nick (1999), "Tradable Permit Systems and Industrial Competitiveness: A Review of Issues and Evidence" in OECD (1999b), *Implementing Domestic Tradable Permits for Environmental Protection*. OECD Proceedings, OECD, Paris.

KITAMORI, Kumi (2002), "Domestic GHG Emissions Trading Schemes: Recent Developments and Current Status in Selected OECD Countries" in OECD (2002a), *Implementing Domestic Tradable Permits: Recent Developments and Future Challenges*, OECD, Paris.

KNIGGE, Markus and Benjamin GÖRLACH (2005a), *Auswirkungen der Ökologischen Steuerreform auf private Haushalte* (Impacts of the ecological tax reform on private households). Report prepared for the Federal Environment Agency, Berlin. (In German.) Available at *www.umweltbundesamt.org/fpdf-l/2810.pdf*.

KNIGGE, Markus and Benjamin GÖRLACH (2005b), *Auswirkungen der Ökologischen Steuerreform auf private Unternehmen* (Impacts of the ecological tax reform on private households). Report prepared for the Federal Environment Agency, Berlin. (In German.) Available at *www.umweltbundesamt.org/fpdf-l/2811.pdf*.

KOHLHAAS, Michael *et al.* (2004), *Economic, Environmental and International Trade Effects of the EU Directive on Energy Tax Harmonization.* Discussion Paper 462, German Institute for Economic Research, DIW Berlin. Available at *www.diw.de/deutsch/produkte/publikationen/diskussionspapiere/docs/papers/dp462.pdf.*

KOHLHAAS, Michael (2005), *Gesamtwirtschaftliche Effekte der ökologischen Steuerreform* (Macroconomic effects of the ecological tax reform). Report prepared for the Federal Environment Agency, Berlin. (In German.) Available at *www.umweltbundesamt.org/fpdf-l/2961.pdf.*

KOHLHAAS, Michael and Stephan BACH (2005), *The impact of special provisions in the framework of energy taxes on the environmental effectiveness – The case of Germany.* Presentation at the 6th Annual Global Conference on Environmental Taxation: Issues, Experiences and Potential, Leuven, Belgium, 22-24 September 2005. Available at *www.law.kuleuven.ac.be/imer/Friday%2023.09.2005/Session%20II%20-%20Economic%20issues%20-%2023.09.2005/2.*

KOUVARITAKIS, Nikos *et al.* (2005), *Impacts of energy taxation in the enlarged European Union, evaluation with GEM-E3 Europe.* Study for the European Commission DG TAXUD. Available at *http://europa.eu.int/comm/taxation_customs/resources/documents/taxation/gen_info/economic_analysis/economic_studies/energy_tax_study.pdf.*

LABANDEIRA, Xavier and José M. LABEAGA (1999), "Combining Input-Output Analysis and Micro-Simulation to Assess the Effects of Carbon Taxation on Spanish Households". *Fiscal Studies* (1999) Vol. 20, No. 3, pp. 305-320.

LAMB, Rebecca and Merrin THOMPSON (2005), *Plastic bags policy in Ireland and Australia.* Briefing prepared for the Scottish Parliament Information Centre. The Scottish Parliament, Edinburgh. Available at *www.scottish.parliament.uk/business/research/briefings-05/SB05-53.pdf.*

LARSEN, Hans (2004), *The Use of Green Taxes in Denmark for the Control of the Aquatic Environment.* Paper presented at OECD workshop on evaluating agri-environmental policies, Paris, 6-8 December 2004.

LEE, D.R., and W.S. MISIOLEK (1986), "Substituting Pollution Taxation for General Taxation: Some Implications for Efficiency in Pollution Taxation", *Journal of Environmental Economics and Management*, 13, pp. 338-47.

Litter Monitoring Body (2004), *System Results – August 2004*, Survey prepared for The Department of the Environment, Heritage and Local Government, Dublin. Available at *www.litter.ie/Litter%20Reports%20August%202004/Systems%20Results%20Annual%20Report%20August%202004%20Final.pdf.*

LIU, Gang (2004), *Estimating Energy Demand Elasticities for OECD Countries. A Dynamic Panel Data Approach.* Discussion Paper 373, Statistics Norway, Oslo. Available at *www.ssb.no/publikasjoner/DP/pdf/dp373.pdf.*

LUTZ, C., B. MEYER, C. NATHANI and J. SCHLEICH (2002), *Innovations and Emissions – A New Modelling Approach for the German Steel Industry.* Paper prepared for the 8th International Conference of the Society of Computational Economics, Aix-en-Provence, France, June 2002.

MÆSTAD, Ottar (2002), *Environmental policy in the steel industry: Using economic instruments.* Presentation made 21 November 2002 for OECD's Joint Meeting of Tax and Environment Experts.

MAJOCCHI, Alberto and Marco MISSAGLIA (2002), *Environmental Taxes and Border Tax Adjustments: An Economic Assessment*, Societa Italiana di Economia Pubblica, Working Papers No. 127/2002.

MARTINSEN, Torhild H. and Erik VASSNES, (2004), "Waste tax in Norway", in OECD (2004a), *Addressing the Economics of Waste*, OECD, Paris. Available at *http://www1.oecd.org/publications/e-book/9704031E.PDF.*

MÄLER, Karl-Göran and Jeffrey VINCENT (eds.) (2001), *Handbook of Environmental Economics.* North-Holland/Elsevier Science, Amsterdam.

McDONALD, Jan (2005), "Environmental Taxes and International Competitiveness: Do WTO rules constrain policy choices?", in ASHIABOR, Hope, Kurt DEKETELAERE, Larry KREISER, and Janet MILNE (eds.) (2005), *Critical Issues in Environmental Taxation: International and Comparative Perspectives: Vol. II*, Richmond Law and Tax Ltd, Richmond, UK 2005, p. 273.

McKITRICK R.R. (1998), "The Econometric Critique of Computable General Equilibrium Modelling: the Role of Functional Forms", *Economic Modelling*, Vol. 15, pp. 543-573.

MOHAI, P. and B. BRYANT (1992a), "Environmental racism". In MOHAI and BRYANT (1992b) (eds.), *Race and the Incidence of Environmental Hazards: A Time for Discourse.* Boulder, Westview Press.

MOHAI, P. and B. BRYANT (1992b) (eds.), *Race and the Incidence of Environmental Hazards: A Time for Discourse.* Boulder, Westview Press.

de MUIZON, Gildas and Matthieu GLACHANT (2004), "The UK's Climate Change Levy Agreements: combining voluntary agreements with tax and emission trading" in Andrea BARANZINI and Philippe THALMANN (eds.), *Voluntary Approaches in Climate Policy*. Edward Elgar, Cheltenham.

MUÑOS PIÑA, Carlos (2004), "Effects of an environmental tax on pesticides in Mexico". *UNEP Industry and Environment*, April-September 2004. Available at *www.uneptie.org/division/media/review/vol27no2-3/530904_UNEP_BD.pdf*.

Ministry of Taxation (2002), "Afgifter og eksterne effekter" (Taxes and external effects) (in Danish) in *SkaÄ – April 2002* (Taxes – April 2002), a periodical issued by the Danish Ministry of Taxation, Copenhagen. Available at *www.skm.dk/publikationer/skat/1501/?crumbs*.

Naturvårdsverket (2003), *Kväveoxidavgiften – ett effektivt styrmedel. Utvärdering av NO$_X$-avgiften.* (Reducing NO$_X$ Emissions. An Evaluation of the Nitrogen Oxide Charge.) In Swedish, but includes an extended summary in English. Swedish Environment Protection Agency, Stockholm. Available at *www.naturvardsverket.se/bokhandeln/pdf/620-5335-3.pdf*.

NEMETZ, Peter N. (ed.) (forthcoming), *Sustainable Resource Management: Reality or Illusion?* Edward Elgar.

NESBAKKEN, R. (1998), *Price sensitivity of residential energy consumption in Norway*. Discussion Paper 232, Statistics Norway, Oslo. Available at *www.ssb.no/publikasjoner/DP/pdf/dp232.pdf*.

NEWBERY, David Michael (2005), "Road User and Congestion Charges" in Sijbren CNOSSEN (ed.) (2005) *Theory and Practice of Excise Taxation*, Oxford University Press, Oxford, United Kingdom.

Nordic Council of Ministers (1999), *The Scope for Nordic Co-ordination of Economic Instruments in Environmental Policy*, TemaNord 1999:550, Nordic Council of Ministers, Copenhagen.

Nordic Council of Ministers (2001), *An Evaluation of the Impact of Green Taxes in the Nordic Countries*, TemaNord 2000:561, Nordic Council of Ministers, Copenhagen. Available at *www.norden.org/pub/ebook/2001-566.pdf*.

Nordic Council of Ministers (2002), *The Use of Economic Instruments in Nordic Environmental Policy 1999-2001*, TemaNord 2002:581, Nordic Council of Ministers, Copenhagen. Available at *www.norden.org/pub/ebook/2002-581.pdf*.

NUTEK (2005), *Näringslivets administrative bördor – Fyra punktskattor* (Industry's administrative burdens – Four excise taxes). Swedish Agency for Economic and Regional Growth, Stockholm (In Swedish). Available at *http://fm2.nutek.se/forlag/pdf/r_2005_07.pdf*.

OECD (1972), *Recommandation of the Council of 26th May 1972 on Guiding Principles Concerning International Economic Aspects of Environmental Policies*. OECD, Paris.

OECD (1974), *Recommandation of the Council of 14th November 1974 on the Implementation of the Polluter-Pays Principle*. OECD, Paris.

OECD (1992a), *Climate Change: Designing a Practical System*. OECD, Paris.

OECD (1992b), *Technology and the Economy: the Key Relationships*, OECD, Paris.

OECD (1993), *Environmental Policies and Industrial Competitiveness*, OECD, Paris.

OECD (1994), *Environmental Taxes and Border Tax Adjustments*, Environment Policy Committee and Committee on Fiscal Affairs, Joint Sessions on Taxation and Environment, COM/ENV/EPOC/DAFFE/CFA(94)31. OECD, Paris.

OECD (1995), *Trade and Principles*, OCDE/GD((5)141, OECD, Paris.

OECD (1996), *Implementation Strategies for Environmental Taxes*, OECD, Paris.

OECD (1997), *Economic/Financial Instruments: Taxation (i.e. Carbon/Energy)*. Annex I Expert Group on the United Nations Framework Convention on Climate Change, Working Paper No. 4, OCDE/GD(97)188. OECD, Paris.

OECD (1999a), *Domestic Tradable Permit Systems for Environmental Management: Issues and Challenges*. OECD, Paris.

OECD (1999b), *Implementing Domestic Tradable Permits for Environmental Protection*. OECD Proceedings, OECD, Paris.

OECD (2000a), *Behavioural responses to environmentally-related taxes*, OECD, Paris. Available at *www.olis.oecd.org/olis/1999doc.nsf/LinkTo/com-env-epoc-daffe-cfa(99)111-final*.

OECD/IEA (2000), *Emission baselines: estimating the unknown*. OECD, Paris.

OECD (2001a), *Environmentally Related Taxation in OECD Countries: Issues and Strategies*. OECD, Paris.

OECD (2001b), *Environmental Indicators for Agriculture; Methods and Results. Volume 3.* OECD, Paris.

OECD (2002a), *Implementing Domestic Tradable Permits: Recent Developments and Future Challenges.* OECD, Paris.

OECD (2002b), *Implementing Environmental Fiscal Reform: Income distribution and Sectoral Competitiveness issues.* Proceedings of a Conference held in Berlin, Germany, 27 June 2002. Available at *www.olis.oecd.org/olis/2002doc.nsf/LinkTo/COM-ENV-EPOC-DAFFE-CFA(2002)76-FINAL.*

OECD (2003a), *Voluntary Approaches: Two Danish Cases. The Danish agreement on industrial energy efficiency, with examples from the paper sector and the milk condensing sector.* OECD, Paris. Available at *www.olis.oecd.org/olis/2002doc.nsf/LinkTo/env-epoc-wpnep(2002)13-final.*

OECD (2003b), *The Use of Tradable Permits in Combination with Other Environmental Policy Instruments,* OECD, Paris. Available at *www.oecd.org/env/taxes.*

OECD (2003c), *Voluntary Approaches for Environmental Policy: Effectiveness, Efficiency and Usage in Policy Mixes.* OECD, Paris.

OECD (2003d), *Environmental policy in the steel industry: Using economic instruments.* OECD, Paris. Available at *www.olis.oecd.org/olis/2002doc.nsf/LinkTo/com-env-epoc-daffe-cfa(2002)68-final.*

OECD (2003e), *Environmentally Harmful Subsidies: Policy Issues and Challenges,* OECD. Paris.

OECD (2003f), *Environmental Taxes and Competitiveness: An Overview of Issues, Policy Options, and Research Needs.* OECD, Paris. Available at *www.olis.oecd.org/olis/2001doc.nsf/LinkTo/com-env-epoc-daffe-cfa(2001)90-final.*

OECD (2004a), *Addressing the Economics of Waste,* OECD, Paris. Available at *http://www1.oecd.org/publications/e-book/9704031E.PDF.*

OECD (2004b), *Sustainable Development in OECD Countries: Getting the Policies Right.* OECD, Paris. Available at *www.oecd.org/document/8/0,2340,en_2649_37425_35112904_1_1_1_37425,00.html.*

OECD (2004c), *Environment and Employment: An Assessment.* Available at *www.oecd.org/dataoecd/13/44/31951962.pdf.*

OECD (2005a), *The Window of Opportunity: How the Obstacles to the Introduction of the Swiss Heavy Goods Vehicle Fee Have Been Overcome.* OECD, Paris. Available at *www.oecd.org/env/taxes.*

OECD (2005b), *The United Kingdom Climate Change Levy: A Study in Political Economy,* OECD, Paris. Available at *www.oecd.org/env/taxes.*

OECD (2005c), *Environmentally Harmful Subsidies: Challenges for Reform.* OECD, Paris.

OECD (2005d), *Manure Policy and MINAS: Regulating Nitrogen and Phosphorus Surpluses in Agriculture of the Netherlands.* OECD, Paris. Available at *www.oecd.org/env/taxes.*

OECD (2005e), *Environmental Fiscal Reform for Poverty Reduction.* DAC Guidelines and Reference Series. OECD, Paris. Available at *www.oecd.org/dataoecd/14/25/34996292.pdf.*

OECD (2005f), *The competitiveness impact of CO_2 emissions reduction in the cement sector.* OECD, Paris. Available at *www.oecd.org/env/taxes.*

OECD (2005g), *The political economy of the Norwegian aviation fuel tax,* OECD, Paris. Available at *www.oecd.org/env/taxes.*

OECD (2005h), *OECD Environmental Performance Reviews: France.* OECD, Paris.

OECD (2006a), *Impacts of Unit-based Waste Collection Charges.* OECD, Paris. Available at *www.oecd.org/env/waste.*

OECD (2006b), Subsidy Reform and Sustainable Development: Economic, Environmental and Social Aspects. OECD, Paris.

OECD (forthcoming), *Instrument mixes for environmental policy.* OECD, Paris.

ORZECHOWSKI, William P. (2001), *Border Tax Adjustments and Fundamental Tax Reform.* Tax Foundation, Background Paper No. 39, November 2001.

Oxford Research (2005), *Måling av næringslivets regelverkskostnader knyttet til særavgiftene.* (Quantification of businesses' administrative burden in relation to excise taxes). Report prepared for the Ministry of Trade and Industry. Oxford Research, Kristiansand. (In Norwegian.) Available at *http://odin.dep.no/filarkiv/257844/Særavgifter-Måling%20av%20administrative%20byrder%20for%20bedriftene.pdf.*

PARRY, Ian W.H. (2005), *Should Fuel Taxes Be Scrapped in Favor of Per-Mile Charges?* Discussion Paper 05-36, Resources for the Future, Washington DC. Available at *www.rff.org/rff/Documents/RFF-DP-05-36.pdf*.

PARRY, Ian W.H. and Kenneth A. SMALL (2005), "Does Britain or the United States Have the Right Gasoline Tax?", *American Economic Review*, Vol. 95, No. 4. A previous version is available as Discussion Paper 02-12, Resources for the Future, Washington DC at *www.rff.org/Documents/RFF-DP-02-12.pdf*.

PARRY, I. W. H., R. C. WILLIAMS III and L. H. GOULDER (1999), "When Can Carbon Abatement Policies Increase Welfare? The Fundamental Role of Distorted Factor Markets", *Journal of Environmental Economics and Management* 37, pp. 52-84.

PEARCE, David W. (1991), "The role of carbon taxes in adjusting to global warming", *Economic Journal*, Vol. 101, pp. 938-948.

PEARSON, Mark (1992), "Equity issues and carbon taxes" in OECD (1992a) *Climate Change: Designing a Practical System.* OECD, Paris.

PESARAN M.H. and R. SMITH (1995), "Alternative Approaches to Estimating Long-run Energy Demand Elasticities – An Application to Asian Developing Countries". In T. BARKER, P. EKINS and N. JOHNSTONE (eds.) (1995), *Global Warming and Energy Demand*, Routledge, 1995.

PETRAS (Policies for Ecological Tax Reform: Assessment of Social Responses) (2002), *Environmental Tax Reform: What Does Europe Think?* A Framework 5 Project: EVGI-CT-1999-0004. AKF, Institute of Local Government Studies, Denmark, CERNA, Ecole des Mines, Paris, Environmental Institute, University College Dublin, University of Surrey, Guildford, Wuppertal Institute, Germany. Available at *www.soc.surrey.ac.uk/petras/reports/european%20policy%20brief.pdf*.

PITSCHAS, Christian (1994-5), "GATT/WTO Rules for Border Tax Adjustment and the Proposed European Directive Introducing a Tax on Carbon Dioxide Emissions and Energy", *Georgia Journal of International and Comparative Law*, 24 (1994-5), 479.

PORTER, M. and C. van der LINDE (1995), "Toward a new conception of the environment-competitiveness relationship"*Journal of Economic Perspectives*, Vol. 9, No. 4.

PORTER, S.J. (1992), "The Tuna-Dolphin Controversy: Can the GATT become environmentally friendly?"*Georgetown International Law Review* V (1992): 91-116.

POTERBA, J.M. (1991) "Tax Policy to Combat Global warming: On Designing a Carbon Tax" in R. DORNBUSH and J.M. POTERBA (eds.) *Global Warming: Economic Policy Responses to Global Warming*, pp. 71-98, Cambridge Mass.: The MIT Press.

QUIRION, Philippe and Jean-Charles HOURCADE (2004), *Does the CO_2 emission trading directive threaten the competitiveness of European industry? Quantification and comparison to exchange rates fluctuations*, EAERE Annual Conference, June, Budapest.

RASPILLER, S. and N. RIEDINGER (2004); *Do environmental regulations influence the location behavior of French firms?*, EAERE Annual Conference, June, Budapest.

RIEDINGER, Nicolas (2005), *Challenges and obstacles in French environmental taxation: recent developments.* Presentation made at the Workshop for Practitioners of Environmental Taxes and Charges, Vancouver, Canada, 17-18 March 2005.

RIVM (2002), *MINAS and Environment, Balance and Outlook.* RIVM, Bilthoven, The Netherlands (in Dutch).

RIVM (2004), *Minerals better adjusted, Fact-finding study of the effectiveness of the Manure Act.* RIVM, Bilthoven, The Netherlands (in Dutch).

ROBERTS, Marc J. and Michael SPENCE (1976). "Effluent Charges and Licenses under Uncertainty" in *Journal of Public Economics*, Vol. 5, pp. 193-208.

RØED LARSEN, Erling (2004), *Distributional Effects of Environmental Taxes on Transportation – Evidence from Engel Curves in the United States*, Discussion Papers No. 428, Statistics Norway. Available at *www.ssb.no/publikasjoner/DP/pdf/dp428.pdf*.

SCHLEGELMILCH, Kai (2003), *Overcoming the Income Dirstibution Obstacle – Distributive Effects of the Ecological Tax Reform in Germany.* Paper prepared for a meeting of OECD's Joint Meetings of Tax and Environment Experts.

SCHLEICH, J. and R. BETZ (2005), *Incentives for energy efficiency and innovation in the European Emission Trading System*, ECEEE Summer Study, Mandelieu, France, Available at *www.eceee.org*.

SCHOU, Jørgen (2005), *The Danish Pesticide Tax.* Presentation made at the Workshop for Practitioners of Environmental Taxes and Charges, Vancouver, Canada, 17-18 March 2005.

SCHÖB, Ronnie (2003); *The Double Dividend Hypothesis of Environmental Taxes: A Survey*. FEEM Working Paper No. 60.2003; CESifo Working Paper Series No. 946. Available at *http://ideas.repec.org/p/ces/ceswps/_946.html*.

SCHRÖDER, J.J., H.F.M. AARTS, H.F.M. ten BERGE, H. van KEULEN and J.J. NEETESON (2003), "An evaluation of whole-farm nitrogen balances and related indices for efficient nitrogen use". *European Journal of Agronomy*, 20, pp. 33-44.

SCHWERMER, Sylvia (2003), *Distributional Effects of Environmental Policy in Germany*. Contribution prepared by Germany for the OECD Workshop on the Distribution of Benefits and Costs of Environmental Policies: Analysis, Evidence and Policy Issues, Paris 4-5 March 2003.

SERRET, Ysé and Nick JOHNSTONE (eds.) (2006), *The Distributional Effects of Environmental Policy*, Edward Elgar, Cheltenham and OECD, Paris.

SFT (Statens Forurensningstilsyn) (2001), *Reduksjon av SO_2-utsleppa i Norge* (Reduction of SO_2-emissions in Norway), SFT-Rapport 1814/2001 from the Norwegian Pollution Control Authority. Available (in Norwegian) at *www.sft.no/english/publications*.

SHAFFER, G. (1998), "The WTO shrimp-turtle case", *International Trade Reporter* 15 (1998) 7: 294-301.

SMITH, Stephen (1998), *Distributional Incidence of Environmental Taxes on Energy and Carbon: a Review of Policy Issues* Paper presented at the colloquy of the Ministry of the Environment and Regional Planning, "Green Tax Reform and Economic Instruments for International Cooperation: the Post-Kyoto Context", Toulouse, 13 May 1998.

SORRELL, S (2002), *The Climate Confusion: Implications of the EU Emissions Trading Directive for the UK Climate Change Levy and Climate Change Agreements*. Brighton: Science Policy Research Unit, University of Sussex.

STAVINS, R. N. (2001), "Experience with Market-Based Environmental Policy Instruments" in Karl-Göran MÄLER and Jeffrey VINCENT (eds.) (2001), *Handbook of Environmental Economics*, North-Holland/ Elsevier Science, Amsterdam.

STERNER, Thomas (2003), *Policy Instruments for Environmental and Natural Resource Management*. Washington DC: Resources for the Future, The World Bank and Swedish International Development Agency.

STERNER, Thomas and Lena HÖGLUND ISAKSSON (2006), "Refunded emission payments theory, distribution of costs, and Swedish experience of NO_x abatement". *Ecological Economics*, 57, 93-106.

Swedish Environmental Protection Agency (1997), *Environmental Taxes in Sweden*. Swedish Environmental Protection Agency, Stockholm.

Swedish Environmental Protection Agency (2000), *The Swedish charge on nitrogen oxides – Cost-effective emission reduction*. Swedish Environmental Protection Agency, Stockholm. Available at *www.internat.naturvardsverket.se/documents/pollutants/nox/nox.pdf*.

SYMONS, Elizabeth, John PROOPS and Philip GAY (1994), "Carbon taxes, Consumer Demand, and Carbon Dioxide Emissions: A Simulation analysis for the UK". *Fiscal Studies* (1994), Vol. 15, No. 2, pp. 19-43.

THALMANN, Philippe (2004), "The public acceptance of green taxes: 2 million voters express their opinion". *Public Choice* 119: pp. 179-217, 2004. Available at *http://reme.epfl.ch/webdav/site/reme/shared/The%20public%20acceptance%20of%20green%20taxes*.

TIETENBERG, Tom (1998), *Tradable Permits and the Control of Air Pollution in the United States*. Working Paper, Colby College, Department of Economics, Waterville, Maine. Available at *www.colby.edu/personal/t/thtieten/permits.pdf*.

TINBERGEN, Jan (1952), *On the Theory of Economic Policy*, North-Holland Publishing Company, Amsterdam.

TIEZZI, Silvia (2001), *The Welfare Effects of Carbon Taxation in Italian Households*. Working Paper 337, Dipartimento di Economica Politica, Universita degli Studi di Siena. Available at *www.econ-pol.unisi.it/quaderni/337.pdf*.

UBA (Umweltbundesamt – Federal Environment Agency) (2004), *Quantifizierung der Effekte der Ökologischen Steuerreform auf Umwelt, Beschäftigung und Innovation* (Quantification of the effects of the ecological tax reform on the environment, employment and innovation), Umweltbundesamt, Berlin. (In German.) Available at *www.umweltdaten.de/uba-info-presse/hintergrund/oekosteuer.pdf*.

UNEP (2001), *Illegal Trade in Ozone Depleting Substances: is there a hole in the Montreal Protocol?* OzonAction Newsletter Special Supplement (number 6), available at *www.unep.org/ozone.*

UNEP (2002), *Production and Consumption of Ozone Depleting Substances under the Montreal Protocol: 1986-2000,* available at *www.unep.org/ozone.*

WEST, Sarah E. and Robertson C. WILLIAMS III (2004), "Estimates from a Consumer Demand System: Implications for the incidence of environmental taxes", *Journal of Environmental Economics and Management,* 47 (2004), pp. 535-558.

WTO (1996a), *United States – Import Prohibition of Certain Shrimp and Shrimp Products, Request for Consultations by India, Malaysia, Pakistan and Thailand,* Geneva: WTO, WT/DS58/1, available at *www.wto.org.*

WTO (1996b), *Japan – Taxes on Alcoholic Beverages,* Report of the Panel, WT/DS8/R, WT/DS10/R and WT/DS11/R, adopted 11 July 1996 (96-2651).

WTO (1996c), *Japan – Taxes on Alcoholic Beverages,* Appellate Body Report, AB-1996-2, WT/DS8/AB/R, WT/DS10/AB/R, and WT/DS11/AB/R. adopted by the Dispute Settlement Body, 4 October, 1996.

WTO (1996d), *United States – Standards for Reformulated and Conventional Gasoline,* Report of the Appellate Body, WT/DS2/AB/R, adopted 20 May 1996.

WTO (1997a), *Canada – Certain Measures Concerning Periodicals,* Appellate Body Report (Canada-Periodicals), WT/DS31/AB/R, adopted 30 July 1997.

WTO (1997b), WTO Secretariat Note for the Committee on Trade and the Environment, *Taxes and Charges for Environmental Purposes – Border Tax Adjustment,* WTO Document WT/CTE/W/47, 2 May 1997: 29 (CTE BTA Note).

WTO (1998a), *United States – Import Prohibition of Certain Shrimp and Shrimp Products, Final Report.* Geneva: WTO, WT/DS58/R.

WTO (1998b), *United States – Import Prohibition of Certain Shrimp and Shrimp Products, Notification of an Appeal by the United States.* Geneva: WTO, WT/DS58/11.

WTO (1998b), *United States – Import Prohibition of Certain Shrimp and Shrimp Products, Report of the Appellate Body.* Geneva: WTO, WT/DS58/AB/R.

WTO (2000a), *United States – Import Prohibition of Certain Shrimp and Shrimp Products, Recourse to Article 21.5 by Malaysia.* Geneva: WTO, WT/DS58/17.

WTO (2001c), *United States – Import Prohibition of Certain Shrimp and Shrimp Products, Recourse to Article 21.5 by Malaysia, Report of the Panel.* Geneva: WTO, WT/DS58/RW.

WTO (2002), *Energy Taxation, Subsidies and Incentives in OECD Countries and their Economic and Trade Implications on Developing Countries, in particular Developing Oil Producing and Exporting Countries,* submission by Saudi Arabia to the WTO Committee on Trade and Environment, WT/CTE/W/215, 23 September 2002.

YANDLE, T. and D. BURTON (1996), "Re-examining environmental justice: a statistical analysis of historical hazardous waste landfills siting patterns in metropolitan Texas", *Social Science Quarterly* 77, pp. 477-492.

OECD PUBLICATIONS, 2, rue André-Pascal, 75775 PARIS CEDEX 16
PRINTED IN FRANCE
(97 2006 08 1 P) ISBN 92-64-02552-9 – No. 55121 2006